Choosing and Using a
Dobsonian
Telescope

For Beebs, who always has to help...

Table of Contents

Introduction

Dobsonian telescopes are the top seller across the globe, and for good reason. They provide good views, simple operation, and economical prices. For all of their virtues they hold many secrets that are not revealed though included instruction manuals or even though a few years' experience.

I have spent quite a few nights helping newcomers understand the basics and start to use their new telescope. There have also been a few who made purchasing mistakes and realized it too late to make a difference.

This all led to the desire to write this book and help people new to the hobby pick out their first Dobsonian, as well as to help the new purchaser get as much out of their telescope as possible.

While I have tried to cover as much information as is practical, this book could easily be three times as long and still not cover everything. Indeed, I am not sure any book of any length could ever cover everything. If someone can read this book and have it reduce their frustration and increase their enjoyment of this amazing hobby, then I consider my work a great success.

The book is designed in four main areas of which "What is a Dobsonian?" is the first. Here we will explore exactly what makes a telescope a Dobsonian, and why that matters. I will share a little of the history behind the instrument in the hopes you will gain a better understanding of why things are the way they are. We will then explore many of the major manufacturers and what they may bring to the table that others may not.

The next section is all about selecting a Dobsonian for you. I will discuss the majority of concerns most first time buyers may have such as size, quality, accessories and much more. If you are looking to buy your first one, here is the section you are interested in.

Choosing and Using a Dobsonian Telescope

Now we come to the section on using your new telescope. What good is having a nice new toy if you don't know how to use it? All you have to do is point it up at the sky right? Well, yes and no. There is actually a little bit of setup and adjustment you need to do first and we will cover all of that to make sure you have fun as soon as you get it out under the night sky.

In this section we will also cover things such as tracking, astrophotography (yep, you will learn how to take pictures with your new telescope!) and accessories you may want to pick up to help you enjoy your new hobby even more.

The last area of the book will show you where you can go to get more information, and trust me, there is a lot more to be had, as well as providing a glossary and index.

I hope that this book will not only get you started but also stick with you long enough to help you many years down the road. Above all, I hope you enjoy your time under the stars.

Clear skies!

Allan Hall

Section 1: What is a Dobsonian?

The term "Dobsonian" (commonly referred to as a "Dob") is derived from John Dobson who popularized the basic premise and design of a style of telescope mount, which uses a simple altitude azimuth and Lazy Susan design to provide an inexpensive but excellently performing mount for a reflecting telescope.

The telescope tube is actually a Newtonian design first introduced around 1668 by Sir Issac Newton. What made this design different from previous reflector designs was the inclusion of an eyepiece near the front opening of the telescope with a diagonal mirror inside to allow easy viewing of the image.

Precursors to the Dobsonian base were seen as early as 1941 at the Springfield Telescope Makers Stellafane Observatory convention.

These telescopes are usually fairly fast, meaning that they collect a lot of light and allow you to see some reasonably dim objects for less money than a comparable telescope of other design.

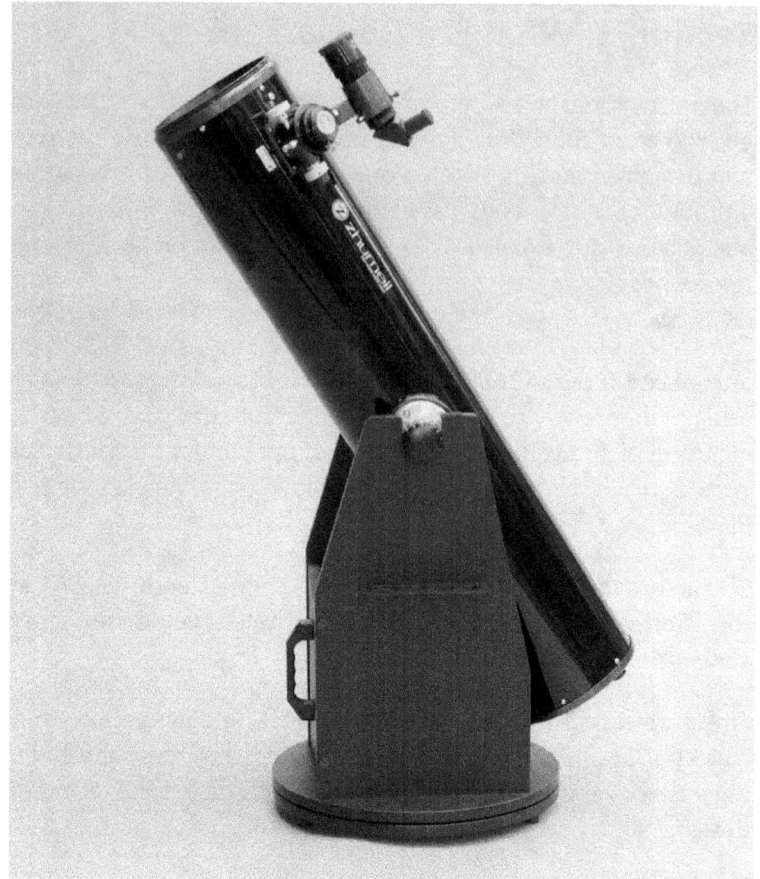

Typical 8" Dobsonian telescope.

These telescopes come in a wide array of sizes, with many different features. You can get Dobsonians that will guide you to your targets, and some that will even point themselves and track targets for you.

If you are just starting out, a good size for beginners would be an 8", or if you have room and can easily transport it, a 10".

There are a lot of different brands and models such as the Zhumell pictured on the previous page, Orion, and Meade. I have a Zhumell that I use for outreach, which is a nice telescope, and I have had excellent customer service from Orion and great luck with their products. Meade makes the

Lightbridge model, which quickly disassembles into something that will fit in all but the smallest cars.

If you want a computerized model, I would probably suggest the Orion push-to models as a great tool to help you find targets quickly.

Since these are Newtonians on a specific type of mount, if you decide to purchase one of these used, pay particular attention to the coatings on the mirrors. These coatings can degrade over time even when the telescope is stored correctly inside a home. Since many people who lose interest in astronomy tend to move these to their attic, garage, or a storage facility, this can substantially increase the rate at which these coatings degrade. This process can take ten or more years so do not worry about one made just a few years ago.

One thing you may want to remember when using a Dobsonian mounted telescope is that the eyepiece can wind up at some extreme places. If you have a small one and are looking at something near the horizon, it is possible that the eyepiece could be just a couple feet off the ground putting you on your knees. Larger scopes when looking directly overhead may require the use of a step stool. Be sure you check the minimum and maximum heights you may use the telescope before you get it in the field so you can have whatever chairs, steps, or kneepads you may need.

Another issue with these telescopes is that they tend to require higher quality eyepieces for the same quality of view. Since these are considered *fast* telescopes, meaning they have a small focal ratio of around f4.5, they show optical imperfections more readily than telescopes with longer focal lengths. This translates into more aberrations and distortions in your view, particularly near the edge of the field of view.

Some people, especially observers with less experience, may not notice these optical distortions. As you gain experience and your observing skill improves, you may start to see these issues and want to acquire higher-end eyepieces. As a very generalist idea, the $25 eyepieces are really bad, the $50 ones are bad, the $100 ones are not too bad, $150 will buy you a pretty nice eyepiece for these telescopes and over $250 will get an excellent one. Remember that those prices are extremely generalized.

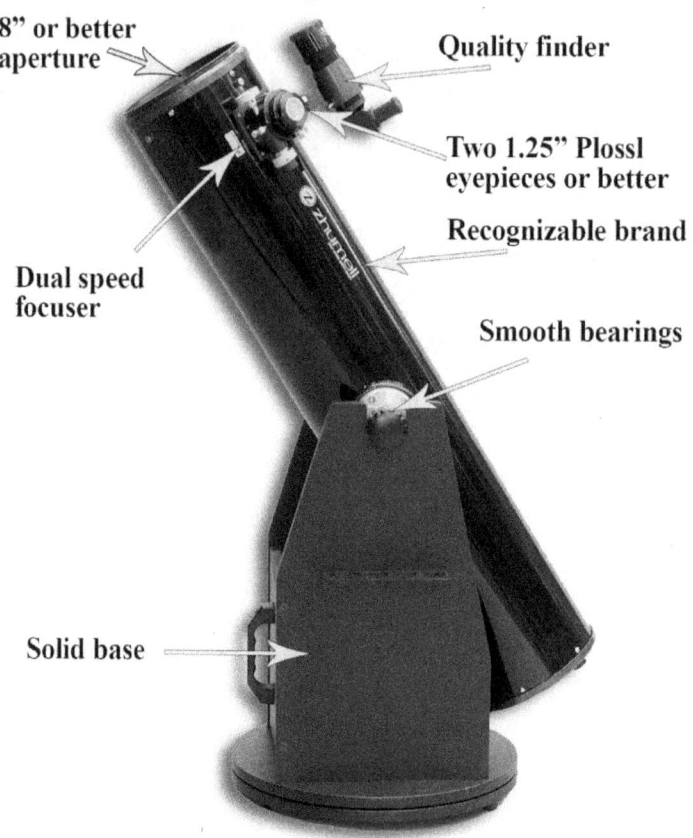

8" or better aperture

Quality finder

Two 1.25" Plossl eyepieces or better

Recognizable brand

Dual speed focuser

Smooth bearings

Solid base

An excellent beginner 8" Dobsonian.

When looking at a used Dobsonian, check the base for any water damage or rot and the mirrors for any problems. Replacement mirrors for these are usually half the cost or more of a new telescope. Bases can be replaced or repaired but you certainly want to know about a problem before it crumbles and sends the telescope crashing to the ground.

Isaac Newton

Newton was born on the 25th of December, 1642 in England to his deceased father (passed away three months prior), Isaac Newton and mother, Hannah Ayscough.

Three years later his mother remarried Rev Barnabas Smith and moved in with him sending Isaac to live with her mother, Margery Ayscough.

After excelling in King's School and moving on to Trinity College, Cambridge, he developed the mathematical theory now called calculus.

While his work continued in mathematics, he turned more and more to optics through the 1670s. This is the period where he built his reflecting telescopes. This work culminated with his publishing of the book Opticks in 1704.

During the late 1670s and through the 1680s his work turned more towards motion and gravity with his law of universal gravitation and laws of motion being published in his book, *Principia* (Mathematical Principles of Natural Philosophy).

Newton passed away on the 20th of March, 1726 in England.

John Lowry Dobson

John Dobson in 2002
(Photo by: Alan J Wylie @ English Wikipedia CC)

John Dobson was one of the most influential people in amateur astronomy of the twentieth century, and most influential in telescope design since Galileo. His "Dobsonian", or dob for short, telescope mount is one of the most popular telescope designs, and one of the best sellers, of all time.

John was born in Beijing, China on the 14th of September in 1915 to his missionary parents, Robert James Dobson (32 years old) and Mabel Annette Lowry (33 years old). He had one older brother Ernest (1914), and two younger brothers Lowry (1919) and Harrison (1923).

Robert was a Zoology lecturer at Peking University in Beijing and Mabel was a musician. It is said that Mabel's father founded the university.

1919 passport photograph of Robert, Mabel, Ernest and John.

In 1927 when John was 12 the family moved from Beijing to San Francisco. A few years later he attended Lowell High School. While it would be a nice story if the school was named after the astronomer Percival Lowell, that is not the case. It was in fact named after the American poet James Russell Lowell.

Dobson was once a member of the Carol Beals dance group. He performed with the group at the Waterfront strike that started because of two worker's deaths on the job. When asked about this performance he said "I was all muscles back then, I had this long hair, and, well, you might understand that I would look a little funny to the dock workers."

After being a "belligerent atheist" since his teens, he attended a lecture given by Hindi Swami Ashokananda, which changed the direction he was to take in life. He returned to school on the urging of the Swami.

In 1943 he graduated from the university with a degree in mathematics and a Master's degree in chemistry. At this time, he worked in the campus lab of Ernest Orlando Lawrence, the Nobel Prize winning nuclear scientist who developed the cyclotron. From there he worked at Caltech and then the Berkeley Radiation Laboratory in areas related to the war.

In 1944 he became a Hindu monk of the Ramakrishna Order at the Vedanta Society monastery in San Francisco. His job at the monastery was to reconcile Hindi beliefs with current scientific theories, which was something that fit his desires perfectly.

Even though the daily duties at the monastery, which started at 4:45am every morning, kept him busy, he still found time to study astronomy. He built his first telescope in 1956, a 12" reflector with a mirror made from the glass of a ship's porthole. Frequently he would sneak out of the monastery to help others with their telescopes and take his to allow other people to get a view of the heavens.

The monastery unfortunately took a dim view of this practice since monks were not allowed to leave without permission, and he never had permission. Eventually this led to an ultimatum being delivered that he cease his telescope making activities or leave the order. Wanting to remain there he did as he was asked.

One night another monk reported him as absent when he in fact was not; this led to his expulsion from the order. It is said that the monks threw his telescope into the San Francisco Bay when he left the monastery.

In the twenty-three years he spent there he managed to cobble together two 18" and fifteen 12" telescopes.

After leaving the monastery, he would set up another 12" telescope he made on the corner of Jackson and Broderick streets in San Francisco on any night with favorable conditions. One of the many people who saw him there helped him find his first astronomy related job and he started teaching Astronomy at the local Jewish Community Center and then at the Lawrence Hall of Science.

In 1968 he started the San Francisco Sidewalk Astronomers club with amateur astronomers Bruce Sams and Jeff Roloff. Throughout the 1970s and 1980s John traveled with the SFSA club all over the western US from one National Park to the next giving demonstrations to anyone interested.

The remainder of his life he spent traveling around the world from one star party to the next, never holding down a steady job, extolling the virtues of the inexpensive "everyman's" telescope we know as the Dobsonian mounted Newtonian, or just a Dob.

In 1991 he published "How and Why to Make a User-Friendly Sidewalk Telescope" (ISBN 0-913399-64-7), in 2004 he published "Beyond Space and Time" (ISBN 0-972805-19-2) and in 2008 he published "The Moon is New" (ISBN 0-981695-20-5).

He appeared twice on the Jonny Carson show.

John Dobson died on January 15, 2014.

Is a Dob right for you?

The next question is always, which telescope is best for you? That is much like asking what kind of vehicle is best. For some people, a pickup is better because they are constantly hauling things. Others may need a minivan to put all the kids in. Still others may need a small, light car for fuel economy.

The old adage of "bigger is always better" or "aperture is king" in visual astronomy is only part of the story. I much prefer my smaller refractors to my larger Dobsonians. The views just seem "better". The key is getting a telescope with the right combination of features that works well for you.

Take into consideration things like how it will fit into your car, how much each piece you will have to carry weighs, how long it will have to cool down before you can observe, and if you need/want something that will find or help you find objects.

I would highly recommend you go to your local astronomy club and try several different telescopes to see what works well for you and fits in your budget.

The single most important thing is that you get a telescope that you will take out and use. If you really want a purple telescope and that is what makes you happy, get the purple telescope. As long as you use it, that is what matters.

Choosing and Using a Dobsonian Telescope

Type	Refractor	Newtonian/ Dobsonian	SCT	MCT
Focal Length	Short-Med	Short-Med	Long	Longest
Cool Down Time	Shortest	Medium	Long	Long
Cost Per Inch	Highest	Lowest	Medium	Med-High
Wind Resistance	Best	Worst	Medium	Medium
Weight	Medium	Lowest	Higher	Highest
Correctors	Flat Field	Coma	Coma	None
Central Obstruction	None	Small	Large	Large
Needs Collimation	No	Yes	Yes	Yes
Best for objects	Large	Large	Small	Very Small
High quality eyepieces more important?	No	Yes	No	No

All of that being said, if you are looking to get the maximum viewing capabilities for the minimum amount of money, don't mind carrying larger and heavier pieces around, and are willing to make a few sacrifices to get those views, then a Dob might be the right scope for you.

What kind of sacrifices you might ask? In the above chart you may find that a Dob has a medium cool down time which means that when taken out of a home at one temperature, it will take time for the mirror to change temperature to match the outside ambient air temperature.

Until the telescope matches temperatures, the views will be pretty poor.

You will also generally have to spend more on eyepieces (once you become a more advanced viewer).

Most importantly (to me that is) is the fact that because of their high wind resistance and lack of backfocus (discussed later) they are extremely poor for astrophotography, which is what I really love to do.

Choosing and Using a Dobsonian Telescope

Most beginners however are not that serious about astrophotography and are very happy with snapping a few quick pics to post on their social media pages. For this, the Dob will work fine.

Every type of telescope and mount has both good points and bad points. For many newcomers to astronomy the Dob provides everything they need with only minor down sides. This is why it is the single most popular beginner telescope made, and has high-end models that even veteran astronomers use with great success.

Manufacturers

There is a wide variety of different manufacturers of Dobs, and each one brings something different to the table. Yes, all 8" Dobs share many of the same specifications and construction techniques but that does not make them all the same.

Some of these companies may not distribute in your neighborhood. Sky-Watcher for example seems to concentrate in Canada and Europe whereas Orion can be a little more difficult to get in Europe than in the United States.

My suggestion is to figure out the size you need, then the general design (solid tube, truss tube, sliding tube) and finally what electronics you need (none or manual, push-to, go-to).

Now take that information and visit the websites of all the manufacturers I list here and see what they have that fits the bill. Compare features and specifications, and do not discount included accessories.

Finally, go to as many astronomy club meetings and star parties as you can and see some of the models in action before you make your final decision.

Orion

I have owned an 8" Orion Dob for a number of years and have had a lot of experience with a wide array of their products and services. While their basic lower end Dobs are nothing to write home about in terms of features, they are robust and come with excellent before and after sales support.

Another nice thing about Orion is that they make some of the largest and most advanced Dobs available via mass market.

Want a 16" truss tube Dob with go-to features? They make one! Sure, Meade makes a 16" too, but not with go-to capabilities.

Yes, you can get larger ones custom built or from companies that have small runs, but those generally cost substantially more than these Orions.

Orion has Dobs ranging from 4.5" up to 16" in a variety of configurations and types.

These Dobs are popular so they tend to show up often in the used market. You can get a solid performing telescope for a really good price by finding one of these used.

Zhumell

This is my current favorite manufacturer as their products come with tons of great features at an astounding price. What they lack is the level of support that Orion provides.

If you are reading this book however, you may not need much support, and you certainly will like their telescopes.

My 8" Zhumell Z8 came with better eyepieces, a better base, better focuser, better finder, and better balance points than my Orion. It also seems to be built almost as well.

If I had to find a problem with it (other than no real manufacturer service) it would be that the screws they use to adjust the position of the finder are plastic. Bump those screws just a little when moving the scope around and they will snap off.

Fortunately, this is an easy fix. I simply replaced the screws with metal versions from Sears.

You can currently get Zhumell Dobs in 8", 10", and 12" sizes.

Another down side is these tend to only be purchased by people who are reasonably serious about astronomy, which means you almost never see one of these on the used market. At least I never see them.

If you are looking for the most bang for your buck, you will not go wrong with the Zhumell Dobs.

Sky-Watcher

Sky-Watcher is a brand created by Synta of Japan in 1999 to sell amateur astronomy equipment. Synta was created around 1980 and has been well received by many amateur astronomy companies.

Predominately sold in Europe and Canada, you will notice many similarities between Sky-Watcher equipment and other manufacturers such as Orion and Meade.

What is interesting is that while Sky-Watcher offers models which are extremely similar to models offered by these other manufacturers (because Synta makes all of it and Synta owns Sky-Watcher), Sky-Watcher also sells products that the others do not.

Sky-Watcher offers solid tube, truss tube, and sliding tube models in manual and push-to variations. These include desktop models, 6" standard and up to 20" models.

While Sky-Watcher used to be hard to get in the US, they are now readily available although from limited suppliers.

I would not hesitate to purchase a Sky-Watcher if it fit in with what I needed and could be had for a competitive price.

Meade

Meade is an old American brand founded by John Diebel in 1972. They started out selling refractors, and then branched out to Newtonians. Several years later, they finally added Dobs to their catalog.

The LightBridge Dobsonian models from Meade are excellent scopes, which disassemble to fit in a much smaller vehicle than solid tubes Dobs. In addition, they are known for having excellent optics for their price points.

Meade makes the LightBridge series available in 10", 12", and 16" sizes.

If you want a Dob that you will be moving around a lot, particularly in a smaller vehicle, the Meade Dobs would be an excellent choice.

Unfortunately there are no models with push-to or go-to technology but for a manual model, these would be awesome.

Astrosystems

Are you looking for a build it yourself kit? Something bigger than 20"? Something completely custom? Try Astrosystems at www.astrosystems.biz.

I will warn you that you do not want to visit this website unless you are prepared to put down some serious money, as in thousands, not hundreds.

If spending money that amounts to the price of a used car on your telescope does not scare you, there are some serious works of art that will last you a lifetime here.

I have included Astrosystems here so you can see the higher end of things to know where you could go. I do not recommend starting out with a telescope from them.

And many more...

There probably are dozens more that make this style of telescope, and quite likely hundreds that have at some point in time.

One of the problems you may run into is that a lot of companies seem to purchase cheap Dobs and rebrand them as their own. This leads to brands that are here one day, and gone the next. Brands like this are difficult to do research on because they have no history.

The best advice I can give you is to stick with a known brand name for your first or second Dob. When you get into spending a lot of money on one, say more than a couple thousand dollars, then you can look at companies such as Astrosystems, Obsession, and New Moon.

Try before you buy: Star parties

The best advice I can give you is to find the closest astronomy club and show up to a meeting or two. Once you find them, find out when their next star party is.

A star party is a gathering of astronomers and astrophotographers who all set up their equipment at the same place and time. These people are used to newcomers coming by and looking, touching (ask first!), and asking lots of questions.

This is the absolute best place to look at different telescopes and get opinions about different makes and models.

Sometimes you need to be a member of the astronomy club to attend the star party, although most welcome non-members. Even if you have to join the club, and even if the club charges dues, it is worth it.

Most of the star parties for the larger club I am a member of contains three or four different Dobs I can play with. The club even has their own Dobs they store on site and you can play with those too! Nowhere else that I know of can you play with that many different telescopes at once and ask questions of people who actually use them, not salespersons.

In the United States, visit https://nightsky.jpl.nasa.gov/ to find clubs and events in your area.

Section 2: Selecting a Dob

This section will cover a lot of ground because not only do we have to discuss the telescope, but also the accessories that may come with it.

While the accessories certainly can give you a head start on getting you observing in a hurry, they should not be as important as the construction of the telescope itself.

A poor quality telescope with ten cheap eyepieces is a far inferior deal to a excellent quality telescope with one reasonable eyepiece. Keep this in mind when looking at not only telescopes, but also kits that may "save you money" by bundling items you may want with the telescope.

You will hear this over and over again throughout this book, nothing beats hands on experience and the best place to get that is your local astronomy club and their star parties.

Size

Size isn't everything in visual astronomy, but it does count for a lot. When selecting a size I strongly urge you to go out and find some amateur astronomers that will let you play with their telescopes before making a decision.

A 12" Dob will allow you to see more objects, and to see all objects usually better, but at a price. I say usually because smaller telescopes actually tend to do better with detailed objects such as the moon, the sun, and planets (think Saturn and Jupiter) than larger ones. The reason is that smaller telescopes cut through the pollution and water vapor in the atmosphere better than larger ones.

The price I talk about in the previous paragraph is not just money. Larger telescopes take longer to equalize temperature so there is more time waiting before you can view.

Larger telescopes also are heavier and more difficult to transport and set up. Like many amateur astronomers I have larger telescopes I take when I plan on doing a lot of viewing, and smaller telescopes called "grab and go" scopes I use if I am only going to be looking for a short while and don't want to expend the time and energy needed for the larger scopes.

You need to find the largest scope that fits your lifting abilities, storage abilities, and financial abilities that still provides the views you want to see.

Optics

Every Dob has two mirrors in it, and then your eyepiece. If you take the eyepiece out of one and put it in another you may still get a different quality of view. Why is that? Mirrors.

Today the mirrors in most Dobs are actually very good. Decades ago the quality in the lower end was so poor anyone could see the difference. Today, it is difficult for anyone without excellent eyesight and years of experience to tell the difference between a $250 and a $500 mass market Dob.

What this means for you is that you can concentrate more on getting good eyepieces (they do make up 33 % of the optical pathway in a Dob) and less on worrying about how good the mirrors are.

Once you have been at this a few years, then you can think about upgrading to a better mirror, or not. I have been doing this a long time and I have never upgraded a mirror.

Focusers

An important consideration with your choice of scopes is the focuser. Many starter scopes come with a cheap 1.25" rack and pinion focuser.

This can be a problem with very widefield eyepieces as you are likely to get some vignetting of the image. Vignetting is where the center is nice and bright but the outside edges are darker, especially the corners.

Another reason this is a problem is that these cheap rack and pinion focusers are generally much harder to get into very sharp focus, as they tend to jerk in little jumps rather than move smoothly.

Lastly these can sometimes include a lot of flimsy plastic components and this is not some place you want a lot of plastic in a telescope. The plastic will wear faster, move less smoothly, and not provide near the stability of a metal focuser.

There are two basic designs of focusers: rack and pinion (above image on the left) and Crayford (mostly in the dual speed variety, above image on the right, note additional controls shows by arrows). While there are advanced rack and pinion designs, they are few and far between. For a little more money you can have the smooth adjustability of a dual speed Crayford. Crayford focusers use very smooth bearings instead of just geared teeth and allow very slight adjustments with the dual speed feature. This makes it possible to really get the focusing exact which can make it much easier to wring out every last bit of detail in an object.

If you already have a telescope with a 1.25" focuser or are about to purchase a telescope without an option for a 2" focuser, most focusers can be replaced after the fact. This can cost somewhere between $75 and $200 depending on the model. You can even spend many hundreds of dollars on very high-end focusers although I doubt you will get any benefit unless you decide to get seriously into astrophotography.

Replacing a focuser on a Dob is an easy process usually requiring the removal of four screws.

Another option to look for on a quality focuser is a tension adjuster, which can increase or decrease the amount of force needed to move the eyepiece in and out. This comes in very handy as some eyepieces can be quite heavy and if there is no tension adjustment the eyepiece weight can cause the focus to change by itself because the focuser can slip. These tension adjustments are normally a knob on the bottom of the focuser.

You can also look for a focus lock knob or screw on the bottom or top of the focuser. This can be useful to lock the focus so it cannot change once you get it adjusted to exactly where you want it to be.

Quality focusers will use a brass compression ring instead of a setscrew to hold the diagonal or eyepiece in the focuser. This provides a more centered optical train and reduces the wear on the eyepiece at the same time. I would never put one of my higher end eyepieces into a focuser that used a setscrew.

Most 2" and larger focusers will come with an adapter to step down to either 2", 1.25" or both.

Focusers larger than 2" are typically used for astrophotography although there is nothing wrong with using them for visual with the correct adapters.

27

Tube or truss?

The two basic types of Dobs are tube type and truss type.

Tube type Dobs are just large fixed tubes. Think of these as something you could fill with water (DO NOT!). They have solid sides and a solid bottom and are the type shown on the cover of this book.

Truss type Dobs (examples shown above, courtesy of ECeDee~commonswiki) usually have a small solid section at the bottom which houses the primary mirror assembly and may or may not have a solid section at the top which holds the secondary mirror and focuser. The part of the telescope which was a long tube is replaced by bars (trusses) that hold the front and rear sections where they need to be.

Tube types are usually less expensive, hold collimation better, and are slightly more durable. Truss types are generally lighter, pack much smaller, and equalize temperature faster.

Accessories

Virtually every telescope you purchase will come with some accessories. This might be a dust cover, or it might be a whole passel of items.

The thing to keep in mind is most manufacturers (and even people reselling their old telescopes) pile on things in the hopes it will make the deal look too good to be true. Often these items are not worth anything or are items you have no use for.

On the other side is that a telescope with a really nice finder and above average focuser is going to be a lot nicer to use than one without these features and so is worth more.

You need to do the research and find out what all the telescope has, and if those items will benefit you.

What follows is a discussion of those accessories to help you on your decision making path.

Finders

I divide all finders into three categories; optical, red dot, and gun sight.

An optical finder is basically a small telescope with less magnification than the primary telescope it is attached to and usually has a set of crosshairs in the center to make for easy aiming. They come in a range of diameters, many different magnifications (and some with no magnification at all), and with or without image correction.

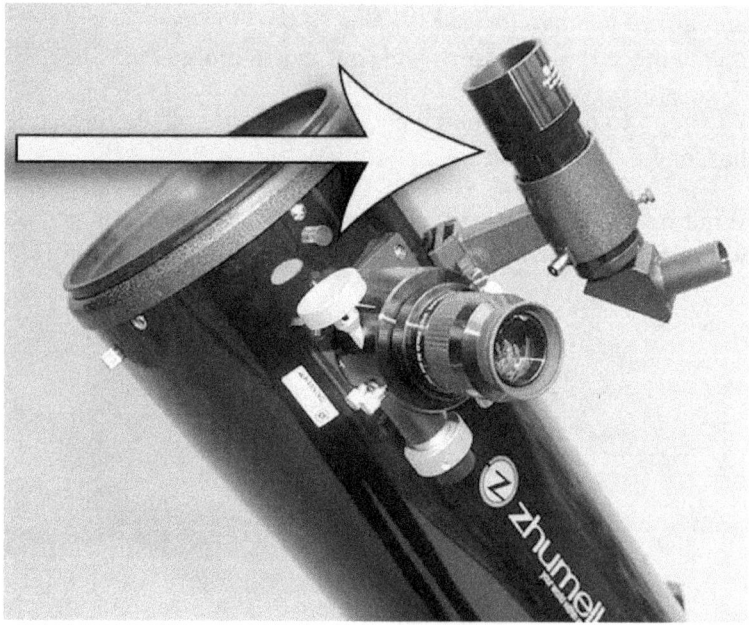

A typical 8x40 optical finder.

The larger the diameter, the more light the finder can gather to make the image it presents brighter. Some finders are so large and gather light so well they are not bad telescopes in their own right. A very nice finder on a typical amateur telescope might have a diameter of up to 50mm while a typical one might be only 30mm or less.

Magnification on a finder is a tricky thing as the greater the magnification, the greater the accuracy. This unfortunately makes it more difficult for roughly finding areas of the sky. A typical range of magnifications of finders would be from about 6x to 9x.

As an image goes through an optical finder, it is reversed and inverted. This can make it difficult to move the telescope while looking through the finder as everything is backwards. To solve this some finders are called correct-image and have a system that reverses the image so that what it shows as up really is up, and what is left really is left. These add a little to the cost of the unit but in my opinion are well worth the price.

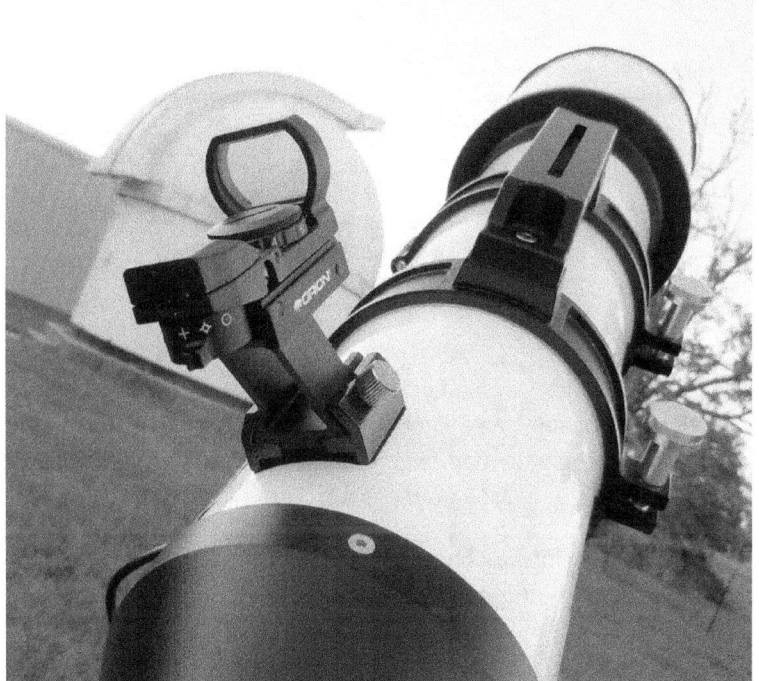

An Orion EZ Finder Deluxe red dot finder.

Red dot finders are a little different in that they have no real optics and no magnification. The basic form of red dot finder is a lit red light on a stick surrounded by a shroud so that as you look down the length of the telescope you see a red dot where your telescope is pointed in the sky.

More advanced models may have a piece of glass or clear plastic with a dot or crosshair projected onto the surface so that the dot appears to float in space where your telescope is pointed.

Many of these have features such as brightness controls, different types of pointers and some even project in different colors (although I would stick with red, as you will learn later).

These are easily my favorite type of finder.

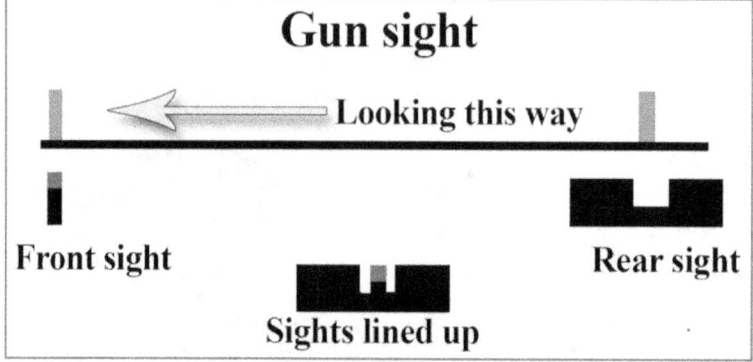

Diagram of a typical gun sight.

Gun sights are very similar to red dot finders except there is no red dot, or usually any lights at all (although the tip of the center sight is usually painted). It is simply a metal or plastic set of pointers that you look down to aim the telescope.

I have even run into a combination gun sight and red dot where the site is designed exactly like a typical gun sight except instead of just a painted tip on the front sight there is a red LED.

Finders are fairly personal and there is no one that is really better than the other. I prefer a nice red dot finder such as the Orion EZ Finder Deluxe and have a difficult time with optical finders.

Eyepieces

Eyepieces are considered half of a telescope's optical system. This makes them quite important. Terrible eyepieces can turn an otherwise serviceable telescope into something you cannot stand to use.

That does not mean you have to spend a fortune on eyepieces to get good ones. It also does not mean you have to have a ton of different eyepieces; a few well-placed focal lengths will serve you just fine.

There are many things to consider with eyepieces including eye relief, apparent field of view, filter threads, coatings and focal length.

Eye relief is the distance your eye needs to be from the glass in order to see the image. This can be quite important for anyone who wears eyeglasses. Even if you do not wear eyeglasses the distance can make it far easier and more comfortable to use. An eyepiece with a small eye relief can be extremely annoying to use. Comfortable eyepieces typically have 12mm-20mm of eye relief.

The apparent field of view (AFOV) is a measure of how much of the sky appears in the eyepiece in degrees. Less expensive eyepieces will have somewhere around 50 degrees or so while higher end eyepieces have 70 or more. Very expensive eyepieces can have up to 120 degrees.

The apparent field of view is not about magnification (how big an object appears). A good example would be that if you were looking at a crater on the moon with two 20mm eyepieces; one with a 55 degree AFOV and one with a 100 degree AFOV. Both would show the crater at the same size in the eyepiece while the 55 degree might show only a small portion of the moon around the crater and the 100 degree would show the entire moon.

Filter threads can be used to screw a filter on the end of the eyepiece. Filters are for a variety of different things including reducing the effects of light pollution and in the case of lunar observing, dimming the image.

Most eyepieces these days have coatings on the glass elements. Coatings are used to reduce stray light and reflections, thus increasing the contrast of your views. You might find eyepieces marketed as coated, multicoated, and fully multicoated, with each being progressively better than the previous.

Focal length is what most people are worried about because this is what gives you your apparent magnification of the object. More is not always better.

Always start out with a focal length somewhere around 20-25mm and then go from there. A very nice set of sizes to start out with would be 25mm, 18mm, 12mm and 8mm. The 8mm would really only be useful during nights of excellent seeing but it never hurts to have one of these for when the seeing does get really good.

To see what your magnification is with a given eyepiece you use the following formula:

Focal length of telescope (mm) / focal length of eyepiece (mm) = magnification

An example would be a 1000mm telescope with a 10mm eyepiece would give us 100x magnification.

Every telescope has a maximum useful magnification which is based off its aperture. The way to calculate the maximum useful magnification is to take the aperture in inches and multiply that times about 30. An example might be a 4-inch telescope multiplied by 30x per inch would be approximately 120x of maximum magnification.

Choosing and Using a Dobsonian Telescope

You will notice in the previous paragraph a lot of vague terms such as maximum *useful* magnification, *about*, and *approximately*. The reason for this is because as magnification increases, image brightness decreases, so there comes a point where detail is lost because it is too dim. More magnification also tends to make things appear less sharp as well. This varies by object and by how sensitive and experienced you are. A very experienced person with very young eyes might get 40x per inch of aperture in magnification out of a telescope while an inexperienced older person might only get 20x.

The maximum useful magnification can also vary based on the quality of the telescope as more light is preserved in better quality telescopes. A high-end APO refractor will present a brighter image to start with and therefore can handle more magnification than a cheap department store reflector of approximately the same aperture.

This maximum useful magnification is important so you do not wind up purchasing eyepieces that provide so much magnification that they will never be useful. An example of this might be for a 4-inch telescope the maximum useful magnification would be around 120x. If the focal length of this telescope was 800mm then you could take 800 divided by 120 to get approximately 7. This means the smallest eyepiece you are likely to be able to use with a 4-inch telescope with a focal length of 800mm would be a 7mm.

Telescope Type	Focal Length	Focal Ratio	Maximum Magnification	Smallest Eyepiece
80mm Short Tube Refractor	400mm	5	90x	5mm
90mm Refractor	900mm	10	102x	9mm
110mm Refractor	770mm	7	123x	7mm
114mm Newtonian	910mm	8	129x	7mm
130mm Newtonian	900mm	6.9	150x	6mm
8" Dobsonian	1200mm	5.9	240x	5mm
10" Dobsonian	1200mm	4.7	300x	4mm
6" SCT	1500mm	10	180x	9mm
8" SCT	2032mm	10	240x	9mm
127mm MCT	1540mm	12.1	144x	11mm
180mm MCT	2700mm	15	204x	14mm

Maximum practical magnification table.

Keep in mind that the figures presented here are estimations of the maximum magnification you could use on the extremely rare occasions when the seeing conditions are perfect. If you observed every night for a year, you might get to use this eyepiece once or twice a year.

You may notice when looking for eyepieces that they come in both 1.25" and 2" varieties (some very old ones may come in smaller sizes such as .965" but these are obsolete and no longer useful). There are two primary reasons for the different sizes: field of view and stability.

Larger sized eyepieces by their nature allow for wider fields of view since the hole that the light passes through is larger. Bigger hole, larger possible image, makes sense. The second reason is that the larger eyepieces tend to be much more stable in the telescope and easier to handle.

If the field of view you want is available in a 1.25" eyepiece and your telescope can handle both size eyepieces, you can opt for the larger eyepiece if you prefer but there is no optical reason to do so. In other words if you want a 30mm Plossl eyepiece and the manufacturer makes both a 1.25" and 2" version, the view will be the same in both of them.

If you have a telescope that uses 2" eyepieces, how can you use a 1.25" eyepiece? With an adapter that comes with most telescopes and diagonals that support 2" or larger items.

Certain eyepieces even fit both a 1.25" and 2" focuser without an adapter as they are made with lands for both. Two such models of eyepieces are the Orion Stratus and Baader Hyperion eyepieces.

When looking for eyepieces there are a few that you should stay clear of. Examples include Huygens which are an old design from the 17th century. These eyepieces suffer from pretty severe aberrations, especially at focal lengths shorter than f10. These type eyepieces are pretty common in very old telescopes and cheap department store telescopes. These eyepieces can usually be identified with an "H", "AH" or "HM" marking.

Ramsden eyepieces are another old design and originated from the 18th century. Very similar in design to the Huygens (both are two element designs) they generally are only slightly better in quality. These eyepieces can usually be identified with an "R", "AR", or "SR" marking.

Kelner eyepieces were another two element design created in the 19th century and are far superior to both Huygens and Ramsdens. Unfortunately these suffer greatly from internal reflections which can substantially reduce contrast and make it harder to see faint objects. These are usually identifiable by the letter "K" marked on them and are another very popular eyepiece included in cheap department store telescopes.

Lastly, you should also stay away from old Erfle eyepieces, specifically ones created around World War II. You can usually find a ton of these at large astronomy gatherings for sale very cheap. These suffer from astigmatisms and internal reflections which make them all but unusable at higher magnifications.

One other type of eyepiece you should be familiar with is a centering eyepiece. These are specially designed eyepieces that have a crosshair inside of them so you can perfectly center a star in the eyepiece. Most are even illuminated to make this much easier by making the crosshairs glow. The purpose of these is to use them when doing a computerized alignment (either using the hand controller on your telescope or an actual computer controlling your telescope). The more accurate your alignment, the more accurate your computer will be at go-to, push-to and tracking operations. To be extremely accurate you do not need one of these with high magnification. The one I typically use is a 20mm model.

When not using your illuminated centering eyepiece, be sure it is turned off and preferably remove the batteries. For some reason they are very easy to turn back on while moving around in your eyepiece case.

When you are just starting out I would recommend a simple selection of reasonable quality Plossl eyepieces from a major supplier such as the Orion Sirius line. Regardless of the type of telescope you have, these eyepieces will give you pretty good views to get your feet wet. These 55° field of view eyepieces tend to run somewhere around $50 each.

For the next step up you could move up to something like the 66° field of view Orion Expanse series. These eyepieces also provide more eye relief than the Plossls and only cost a few dollars more at around $70 each.

From here I would probably suggest saving your pennies and moving to something like the Orion Stratus or Baader Hyperion eyepieces. These two models are similar enough as to be interchangeable for all intents and purposes. If anything, the Hyperions are probably a little newer design and may benefit slightly from newer coatings. I have never actually seen a viewing difference however so whichever you prefer should be fine. These provide a 68° field of view and excellent eye relief. All of them except the largest few Hyperions provide the ability to be used in both 1.25" and 2" focusers without the need for an adapter. Either set of eyepieces are about the same price at around $140 each with the exception of the largest of the Hyperions which are just over $200.

Once you leave the Stratus/Hyperion eyepieces behind, the next real step up is something like the Tele Vue Pantopic/Delos/Nagler/Ethos or Pentax XW. These are among the top eyepieces in the world. Starting at $350 each for the cheapest of them, they should be. Find anyone who is really serious and has been doing astronomy a while and you are very likely to see one or more of these in their eyepiece case. When you want the most perfect view possible, try one of these.

There are of course other brands and models to choose from. What I have mentioned above are the ones I would recommend based off what I have personally seen and played with. There is also no reason to stick with just one brand or model for all your eyepieces. Many astronomers I know mix and match taking the best eyepiece from each line and make their own set.

Barlows

Barlows are a relatively inexpensive way to boost your eyepiece collection, at the expense of a little loss in image quality and object brightness. These devices go between your eyepiece and telescope and come in sizes such as 2x, 3x, 4x, and 5x. If for example you use a 2x barlow on a 20mm eyepiece, it effectively makes your eyepiece a 10mm.

The down side is that this adds a little distortion into the image and reduces the amount of light reaching your eye. How much? That depends partially on the quality of the barlow, the quality of the image to start with, and how good your observing skills are.

I know people who use barlows all the time and see little to no image difference between using it and using a smaller eyepiece. I also know people who cannot stand to use a barlow at all. I personally never use a barlow for visual astronomy although I do own a couple just in case.

When just starting out, barlows such as the Orion 2x shorty or Meade 2x shorty are excellent choices. If you are looking to keep the maximum amount of light and minimum distortion possible, look for something like the Orion three or four element designs or the Tele Vue Powermate series.

Filters

A group of 1.25" visual astronomy filters.

For visual astronomy there are four categories of filters: light reduction, light pollution, nebula, and colored.

Light reduction filters are sometimes marketed as moon filters or polarizing filters and are used to reduce the amount of light reaching your eye from very bright objects, typically the moon. They are not to be used when viewing the sun as that requires specialized equipment. (NOTE: Viewing the sun without the proper equipment can result in blindness, fires, and worse.)

Light reduction filters are not always required on smaller telescopes but become progressively more important as the size of the aperture increases as the telescope will collect more light. Viewing the moon through a good sized telescope without one of these filters can be painful and cause temporary blindness when you look away from the eyepiece. This also explains why the moon filter you start with in a small telescope will not work as well with a larger one.

You may find variable polarizing filters sold as moon filters as well. These are excellent choices as they not only can adjustably reduce the amount of light entering the eyepiece, but also polarize it to prevent extraneous light and internal reflections.

Light pollution filters were mentioned previously in this book so you should already be familiar with them.

Nebula filters are very specialized. When you look at a nebula such as the Orion Nebula you will notice that it is primarily a greenish-gray color. This is because the bulk of the light emitted that we can see is emitted by OIII, or ionized oxygen, and is in the green part of the spectrum. These filters heavily restrict the light passing through and allow only a small portion of the green spectrum of light, right around 500nm where OIII emissions are visible. These can block out substantially more light pollution than a generic light pollution filter and dramatically increase the contrast between a nebula and the black background of space.

Be aware that nebula filters will not work well with nebulae that do not emit in the OIII spectrum. When they do work it will be fairly obvious.

Colored filters are primarily used when viewing planets to help increase the contrast between visible features. I have heard reports that they can also assist in the viewing of nebulae although I have never had them improve my views. There are even colored filters to use with less expensive achromatic refractors which can reduce the chromatic aberrations (blue glow around brighter objects) sometimes found in these telescopes.

Filter #	Color	Use	Notes
8	Yellow	Moon	Can reduce glare and the glowing effect from bright objects when used with an Achromatic refractor.
21	Orange	Mars, Saturn, Jupiter	Increases contrast of surface or cloud features.
23A	Light Red	Venus	Increases contrast between Venus and the daytime sky.
25	Dark Red	Venus, Mars	Increases contrast in cloud features.
38A	Aqua	Mars	Increases contrast in cloud features.
47	Dark Purple	Venus	Increases contrast in cloud features.
56	Light Green	Saturn, Jupiter	Increases contrast in cloud features.
58	Green	Mars, Jupiter	Increases contrast in polar regions.
80A	Light Blue	Mars, Jupiter	Increases contrast in cloud features.

Typical visual filters and their uses.

There are two general places to place filters: screwed on to the nose of the eyepiece and screwed on to the front of a diagonal. If your telescope uses a diagonal I suggest putting the filter there so you can change eyepieces without having to worry about the filter.

I highly recommend you not purchase filters right up front. The overwhelming majority of people never use filters for visual use because the amount of improvement in the image is so minor that it really takes an experienced observer to see it. Once you have been doing this a while, get one that will work on a target you are interested in and play with it. Once you get the hang of it and it makes a difference, buy the next one.

You should also steer away from those kits that have a ton of eyepieces and filters in them. You will never use 90% of that kit and will wind up selling it online for a fraction of what you paid for it.

Cases

Most people do not have a case for their solid tube Dob. There simply is no reason for it.

Truss tube models however very often have cases and can frequently be bought in a kit with telescope and cases already bundled. Orion usually has this option on their truss tube models.

Currently Orion has this option on both their 12" and 14" intelliscope models (push-to computerized telescopes). These same cases fit their non-computerized models as well.

Oddly, they do not have the cases for the 14" available outside the kit, at least on their website. The 12" models are available separately.

Generally speaking, the case is more to make the telescope easier to transport than for protection, although it does offer the latter.

Collimators

All Dobs need to be collimated (have their mirrors aligned). This is not something "done at the factory" that you will not need to do during normal use. This is something you will need to do reasonably often.

There are many devices you can use to collimate your telescope but by far my personal favorite is the laser collimator. This device fits in place of an eyepiece in the focuser and shoots a laser beam down to the primary mirror and back.

While there are devices much cheaper than a good laser collimator, which run between $50 and $100, none of them makes the job as easy.

In fact, I am so much a fan of these things this is the only way I am going to discuss for collimating your telescope.

Sure, you can get a collimating eyepiece adapter for $20-$50, you can also get laser collimators from $25 on up. I would suggest a $25 laser collimator over a $50 collimating eyepiece any day of the week.

Even with that said, as important as this piece is, get something nice at least $60 retail (cheaper used of course).

Other accessories

Telescopes from different manufacturers will come with different accessories, and sometimes the same telescope can come in different kits from the same manufacturer. Sometimes these accessories are important additions, sometimes not.

My Zhumell Z8 for example came with a laser collimator, which saved me around $80. Many models come with star maps and software, which is usually useless.

One exception to this, assuming you are new to astronomy, is the models that come with a Planisphere and the Orion DeepMap 600. I still carry both in my astrophotography kit to this day as they are excellent products to show newcomers to the hobby how things work, and where things are.

Stay away from filter kits (you will never use them), huge eyepiece kits (two to start with is more than enough) and any software (which is always a lite version and way out of date).

You can download free software online such as Stellarium (www.stellarium.org) or apps for your phones/tablets such as Star Walk (www.vitotechnology.com) and Sky Safari (www.skysafariastronomy.com).

47

Push-to and Go-to

Most people start with a manual Dob as their first Dob. This means that if you want to move the telescope, you push it. You have to hunt around in the sky and star hop (find a known star and navigate from star to star using a chart to find an object). But it doesn't have to be that way.

Newer Dobs are available either with electronic packages that not only know where the telescope is pointed, but also can tell you where to push to see what you want or in some cases, move the telescope all by itself and point at your designated target.

Telescopes that tell you where to push are called push-to systems. If the telescope moves to the target by itself, that is a go-to system.

Major manufacturers such as Orion and Sky-Watcher have both types of systems on their telescopes.

It does add a reasonable amount of money to a telescope. Orion's 10" with no electronics runs just over $600, the push-to version is about $900, while the Go-to version is $1500.

Electronics can really make your night more enjoyable because you can find cool things to look at much faster. The down side is that you do not tend to learn the sky as well because the telescope does a lot of work for you.

If my primary telescope was a Dob I would seriously want a Push-to system on it.

While I have several Go-to mounts for my refractors because I do a lot of long exposure astrophotography with them, I do not see spending $600 to go from a Push-to to a Go-to Dob.

Generally speaking you cannot install a computer system on a Dob after purchase so no, you cannot add it later.

Choosing and Using a Dobsonian Telescope

The biggest question you have to ask yourself is, would you use the telescope more, or get more enjoyment out of it, if the telescope told you where to go to see an object?

For me, that answer is usually yes. I do however own a couple of completely manual telescopes that I take out from time to time to have some fun. It can be very rewarding to find your way to a target simply by using a map.

My biggest issue is when I am trying to view a really faint object I can never be sure that I am looking at the right object unless I have some way to verify it.

With my Go-to telescopes I can move to a known object, yep, that's right, move to another known object, yep, looks good, and finally to the object I am not sure about. Must be right.

The important thing to remember here is that there is no best or right way. There is the way you want to do it.

My favorite idea is that whatever makes it more likely that you will go out and use the equipment is what is right. If that means you need to spray paint your telescope tube in fluorescent colors and hang colored beads off the finder, so be it.

Size and weight of components

I cannot stress this enough, go to an astronomy club meeting and/or star party and play with the equipment. Some people are surprised by how much a hollow tube Dobsonian telescope weighs. Not to mention how big the tube actually is.

When I bought my last car I loaded my astronomy equipment into my existing car, drove 45 minutes to the dealership I was looking at new cars at, went and got the salesman, and loaded the prospective new car with my equipment to make sure it fit well.

OK, fine, I was buying a MINI Cooper so I had a little more justification than most people at being concerned about space. I am glad it fit!

If you want a 12" solid tube Dob and you drive a Smart Car, you are in for a sad awakening. Even if you drive a Surburban you may be in for a shock when you reach down and try to lift that 50+ pound optical tube off that 35+ pound base.

Remember, if you hate lifting that huge telescope you are less likely to use it. The views in your 8" Dob out under the stars are a lot better than the views in your 16" Dob that never leaves your garage.

New or used?

This age-old question plagues us all.

New gets us a warranty and that new Dob smell.

Used gets us more bang for our money, with the possibility that something might be wrong with the telescope.

The truth is that as long as you understand the basic workings of a Dob and are not blind, you should be pretty safe on the second-hand market.

The first thing you need to do is make sure there is nothing broken, dented, or cracked. Take a little flashlight and look down the tube to make sure there is not a problem there.

Obvious things are the focuser is not smooth or jerks around. The eyepiece may not be secured in the focuser and may move around. The mirror might be horribly dirty (think covered in something to the point you cannot see a reflection at all. One or all of the spider vanes may be bent.

If you take the telescope out and point it at the top of a tree down the road you should be able to get the tree in focus. If not, something may be wrong. Ask the current owner for assistance. If they cannot find a person who knows how to use it, there is your first clue to run, not walk, away.

I have bought both new, and used, without issue. My Orion Dob was used and has a little issue with the finder (which I replaced almost instantly anyway because I hated it) but that defect was disclosed during the sale.

Personally, if I had access to a used telescope so that I could physically test it before I handed over cash, I have no problem buying used. Buying something over the internet without being able to inspect it first, that is a different story.

Choosing and Using a Dobsonian Telescope

The last time I purchased a used telescope without putting my hands on it before handing over cash was a refractor off of eBay.

From what I can tell, the telescope and mount were in excellent condition when the seller put it in the box. Unfortunately, they did not take into consideration that the weight of objects (such as the counterweights) might cause things to slam together when the delivery person got a little rough.

When the package arrived, a crucial component was broken into several pieces. This component was no longer made and there was no way to use the telescope mount without it. I wound up having to have the piece custom made at a machine shop.

The good news was that after taking some pictures and sending them to the seller, I wound up with a complete refund. That was lucky since the custom piece I had made cost almost as much as I paid on eBay for the telescope to begin with.

The moral of the story here is that buying anything used has risks. You could have a seller who does not know, or does not disclose that the telescope has a major problem. You could have a perfectly honest seller and a great telescope, which is destroyed in shipping.

You can minimize or eliminate a lot of those risks by buying locally. Look on Craigslist and at your local astronomy club. If you are in a large city or a college town, you can also try pawn shops although I am very leery of those.

Even more considerations

There are several other things you may not have thought of, and we will cover some of those here.

If you buy a truss tube Dob you will need a light shroud, these cover the opening between the two sections of the telescope and prevent stray light from entering the light path. This is a necessary item.

When viewing through any telescope, including a Dob, you may need a chair to sit in. Unfortunately, you cannot really use a standard chair because it is not adjustable. Believe me, you will need one that is adjustable.

Many astronomy shops have chairs specific to astronomy. You can also build one such as a Denver Chair as shown below.

Most Dobs these days either come with a fan in the bottom, or have one you can buy as an accessory. These usually run off batteries and help cool the mirror down. Without one, you could be waiting for hours for your telescope mirror to equalize with ambient air temperature so you get good views. Otherwise, your views could be horrible.

Section 3: Using your Dob

So you got a Dob and now you need to use it, no problem! Let us look at what you need to do to get started using it.

First, we will learn a little about setting it up, and then we will go more into how to use it out in the field.

Fortunately for us, both setting it up and using it is pretty straight forward so we will have you out in the field looking at the stars in short order.

Getting it ready

Before we can take our new telescope outside and enjoy the awesome views of the heavens, we need to do a little work. This may vary depending on whether you bought it new or used but much of it will be the same either way.

Without this setup, your telescope may be completely unusable. At the very least, it will be much more difficult to use.

Not to worry because the setup of Dobsonian telescopes is pretty easy. Once you get it put together the rest is more about learning how it works than anything really complicated.

The second time, and every time thereafter, it will get faster and easier.

Some assembly required

I have never purchased a new telescope and had it arrive in a ready-to-use condition. There is always something that has to be assembled. Fortunately assembling a Dob is pretty easy.

When you open the boxes you will probably see one box with a bunch of wood (or wood like) pieces that make up the base and another box, which includes the optical tube. Sometimes a third box includes accessories as well.

If any tools are required for assembly it is usually limited to a couple of screwdrivers. Often fasteners included in these use an allen head but those include the wrench with the kit.

Virtually always there will be assembly instructions included and should they be missing, they are downloadable from the internet at the manufacturer's website.

Generally the base may take fifteen minutes or so to assemble and includes steps such as fastening the bearings to one piece and then sandwiching that to another piece, then building a little box that the telescope tube sits in.

The telescope tube rarely needs any assembly beyond attaching the finder and possibly some pieces on the side that help hold or provide tension to the base.

Although I have been lucky so far, do not expect any batteries for finders or collimators to be included. As soon as you expect them to be there, they won't be.

If you have any problems with the assembly and you purchased from Orion, Meade, or another popular company, you can usually call them for assistance. I had to call Orion one time because a couple screws were missing. I didn't know they were missing, I just couldn't figure out how to attach two things together. They shipped me a new screw set out by priority mail.

Collimation

If you have a Dobsonian telescope you will need to collimate it. Collimation is the process of making sure all the light that is gathered coming in your scope is correctly focused in your eyepiece. Let's start with an overview on a Newtonian:

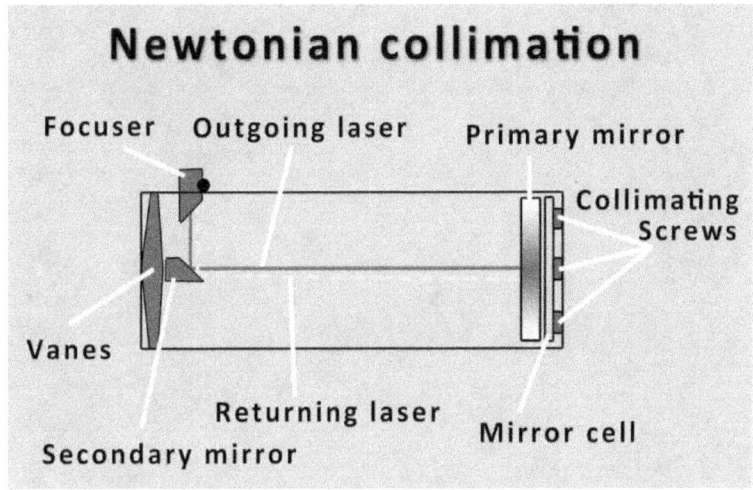

Laser collimation diagram on a Newtonian telescope tube.

The objective here is to make sure that the laser beam from a laser collimator once shined in the focuser comes exactly back into the focuser, which means the mirrors are lined up.

The first step in laser collimating your Newtonian is to make sure the laser is on, inserted and centered in the eyepiece holder of the focuser and then to make sure it is shining in the center of the circle in the middle of your mirror (or on the central dot). In order to do this, you need to look down the telescope tube.

I would recommend putting a piece of paper over the telescope opening after inserting the laser collimator and turning it on. Now look for a dot on the paper. If you see the dot be sure to avoid it when you look down the tube. If you do not see the dot, it is probably safe to carefully look down the tube.

Once you are sure it is safe to do so, look down the end of the tube and you should see the laser dot on the large mirror at the other end of the tube.

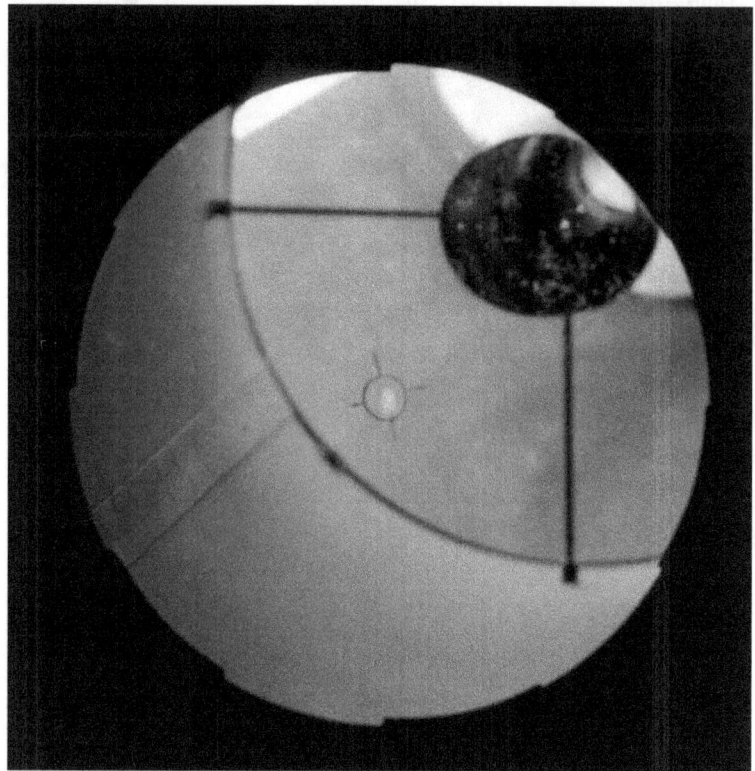

Laser shining in central circle during Newtonian collimation.

You need to get the laser dot right in the center of the circle marked on the mirror (some mirrors have circles, some have dots). If it is not, you need to adjust the secondary using the screws on the front of the telescope opening. These screws are on the back of the secondary mirror suspended by the spider vanes.

Secondary collimation screws.

When adjusting the secondary mirror you need to move the screws in very small motions, about a quarter turn at a time. When you tighten one, loosen the other two. Do this slowly and methodically until you get the laser dot right in the center of the circle or dot on the primary mirror.

Once the secondary is aligned you need to look at the laser collimator as it has a target on it and you should hopefully now see a laser dot on that target.

Collimator target showing returning laser location.

The laser dot should be right on the center dot or crosshairs of the laser collimator's target. If it is not, we need to adjust the primary mirror cell on the back of the scope. Be sure that the target you see in the above image is pointed towards the back of the scope so you can see it while adjusting the primary mirror cell, which should have three screws much like the ones on the secondary mirror we already adjusted.

Rear view of the primary mirror cell and adjustment screws.

Just like we did with the secondary adjustments we need to tighten one screw about a quarter of a turn and then loosen the other two, check the collimator target and continue adjusting if necessary.

61

Finder alignment

The finder needs to be aligned before your first night out or you will have a difficult time finding objects in the sky. Finders have some method of adjusting where they point such as adjustment screws on the finder mount.

A typical optical finder mounted to a telescope with adjustment screws.

This process is far easier before dark. Pick a distant object that is easy to see with your naked eye (I like to use power poles). Using a low power eyepiece (an illuminated centering eyepiece is best) center the object in your telescope and lock the mount so that the telescope cannot move.

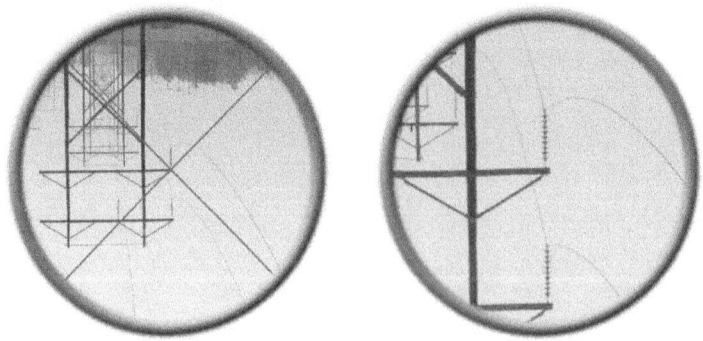

View through the finder on the left, telescope on the right.

Now adjust the finder until the crosshairs (or dot if you are using a red dot finder) is precisely centered where it should be.

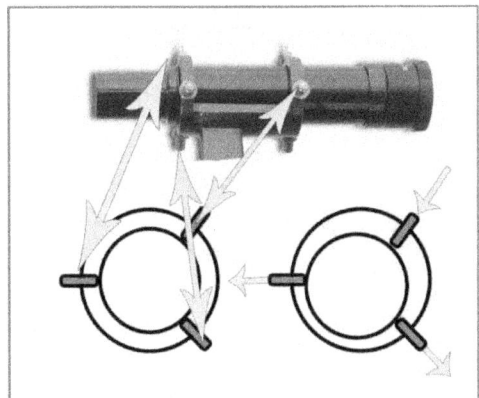

Adjustment screws for an optical finder.

To move the finder there are several methods in use. The primary adjustments for most visual finders are by using setscrews. To move the finder you tighten one screw and loosen another to move it.

Depending on the desired direction of motion you may have to loosen two screws and tighten one, or tighten two and loosen one.

Originally finders used two sets of setscrews: one to adjust the front of the finder and one to adjust the rear. You will also see some designs where the front or rear of the finder is supported by the bracket and the setscrews are on the opposite end. This design is especially popular with the larger optical finder designs.

An optical finder with one set of adjustment screws.

You may also see either of these designs with a spring pin in place of one of the setscrews. This makes it easier to adjust as one of the three setscrews always applies tension against the tube which can be overpowered by the other two if needed.

Other finder types such as red dot finders usually have much different adjustment procedures which still operate on the same basic principles.

Many red dot finders have two setscrews to raise the front or rear up and down, and left and right. They may also have a rail system which can be similarly adjusted.

An Orion EZ Finder Deluxe.

Why are things upside down?

Depending on the type of optical aid and telescope you are using, the image you see may not be what you expect.

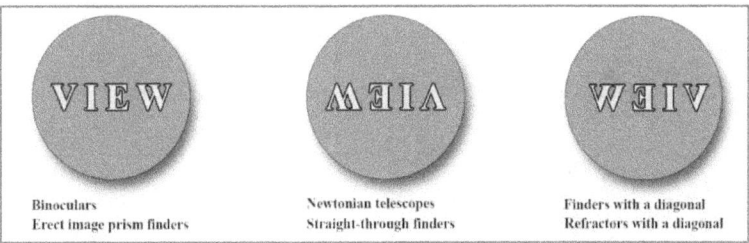

Binoculars
Erect image prism finders

Newtonian telescopes
Straight-through finders

Finders with a diagonal
Refractors with a diagonal

Views differ depending on the design of the device.

Binoculars and finders with erect image prisms in them will show you an image that is correct; that is to say, if you looked at a billboard it would appear as if you were not using an optical aid at all.

Newtonian telescopes, including the Dobsonian mounted Newtonians, and optical finders that are straight-through will render the image upside down and backwards.

Most refractors come with a 90 degree diagonal as do some higher end finders. These will present an image that is backwards but right side up.

Diagonals are made for refractors which use a prism and can show a correct image as well.

Many people wonder why this really matters, who can really tell if a star is right-side up or not anyway? It matters when viewing the moon and comparing that to a map, as well as when you are moving the telescope to try to navigate the sky. Pushing the nose of the scope to the right only to see the image move left takes some getting used to.

Computer alignment

If you have a go-to or push-to telescope, you will need to align the computer. What follows is a typical sequence of events which may not follow your handset's procedure exactly but should be close.

The first thing most mounts want to know is where on the planet you are. Here you will enter your longitude and latitude. If you do not know this information you can get it from most smartphones, GPS units, navigation systems or online with most mapping sites.

Be sure you enter the E/W or N/S identifiers correctly. Failing to enter the correct identifier will completely confuse the mount and cause you to not be able to find any targets.

Also make sure you are entering your longitude and latitude in the correct places. Attempting to reverse the placement of these numbers will not make you a happy person. If the computer even accepts your data, it will never be able to find any targets much less track them.

The next thing you will need to know is your time offset. This little piece of information is the number one mistake I see novices make when it comes to setting up their new telescope.

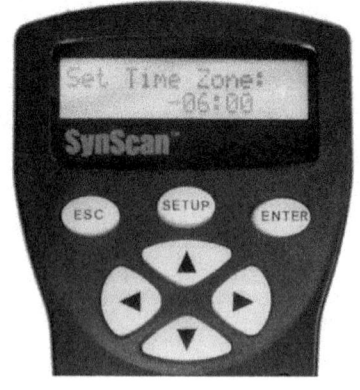

This number is based off of your UTC (Universal Time Coordinates) time (or Zulu time) and is known as a UTC offset. If you do not know your UTC offset you can use the chart on the following page to locate it.

Note that this number does not change based on daylight savings time. It remains the same all year long.

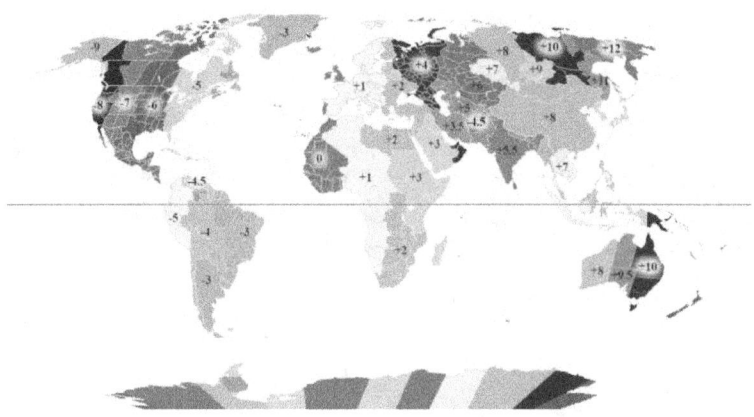

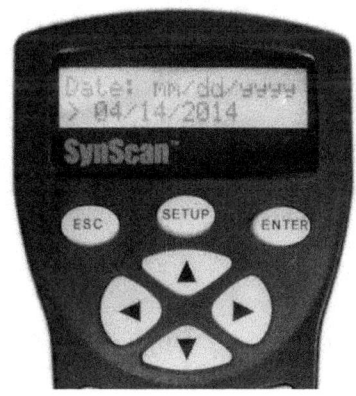

Next we have to enter the date. There is nothing special to remember here except that the date is in U.S. format so if you are across the pond do not forget to reverse the month and day as shown.

After entering the date we need to enter the time. Most hand controllers I have dealt with use twenty four hour time instead of the normal AM and PM twelve hour time. Since you will most likely be out in the evening just add twelve to the time and you should have it.

For example, if it is 7pm just add 12 + 7:00pm = 19:00:00. It is not critical to be exact to the second.

The next question is for daylight savings. This is probably the second biggest mistake I see novices make. If you are on daylight savings time, select yes. If not, enter no.

If you are not sure, you can check online to see. For me here in the central United States in 2017 daylight savings time runs from March 12th to November 5th. Between those two dates I select "yes", after November 5th and before March 12th I select "no".

Here we have a fork in the road. When I am doing my normal visual without a laptop computer connected to my mount I select "1)YES" and allow the handset to do an alignment. When I have my laptop hooked up to the handset and want to use it to control the mount, I select "2)NO".

68

Connecting your computer to your mount is beyond the scope of this book but I did want to mention where the fork occurred just so you would be aware.

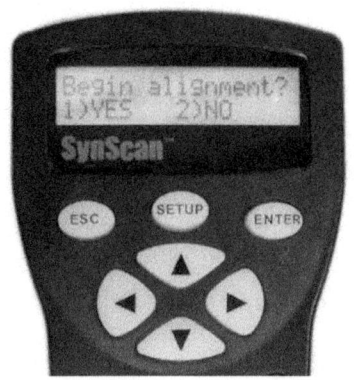

For the purposes of this book, I will assume you selected "1)YES".

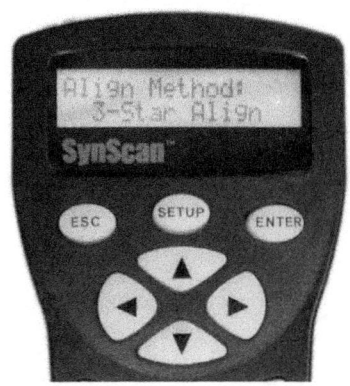

The handset now wants to know which method you want to use to align the mount. There are usually several options: 3-Star Align, 2-Star Align and 1-Star Align.

The most accurate alignment is achieved when you use the three star alignment. When using this method you should use three stars that are widely spaced apart making sure one star is on the opposite side of the meridian from the other two.

Once you are finished with this alignment routine you are set up and ready to start observing!

Out in the field

Some things you can do right in your living room, but others may require you to be outside where you plan on viewing in the dark.

We just finished discussing computer alignment and while most of that can be done in the daytime, the last part where you align the unit based on stars needs to be done at night.

I choose to put that section under "Getting it ready" instead of in this section because I think getting as much of it done as possible in your living room (entering your coordinates, time zone, date, time, etc) the first time will go a lot better if it isn't dark and you have access to the book and internet.

After the first time you will do the entire computer related stuff out in the field.

Let us move on to things you have to do in the field.

Temperature equalization

Pretty much all telescopes make you look through glass by looking through a lens, a mirror, or both. This presents our first problem in that glass tends to expand and contract as it heats or cools (as does everything else in your telescope).

When the shape of the glass (mostly the primary mirror because it is a huge chunk of glass) changes, the focus and collimation changes. This causes things to get blurry.

The second problem is that when you take a warm mirror outside (actually the entire telescope, but again, the huge brick of glass that makes up the primary is the biggest issue) into cool air, heat leaves the mirror into the surrounding air causing air currents.

You can see this effect in the summer when driving down the road and you see things in the distance shimmering as heat leaves the ground.

Both problems happen in reverse as well while the mirror is absorbing heat from the surrounding air such as in the summer.

To get your telescope to equalize as fast as possible, remove any covers or caps to make sure air can flow as good as possible. If your unit has a fan for the primary mirror, turn it on. Now sit back and wait about 10 minutes per inch of aperture. This means just under an hour and a half for an 8" Dob.

To test and see if it is ready, point it at a bright star as close to directly overhead as possible and turn the focuser all the way in or out until your star is a big round blur in the eyepiece. Now look and see if there is any wavy motion in the image (like the heat waves you see in the distance above the road in summer).

If there are waves, it needs to cool longer. No waves means it is ready to go.

Choosing and Using a Dobsonian Telescope

Tube type dobs will generally cool fastest when pointed straight up since the heat can be radiated through the top of the telescope easier than through the side.

I should warn you however that things happen when telescopes are left without their covers on them pointed straight up. Rain, leaves, and bird poop are just a few things I have seen or heard go down the telescope tube. It seems that birds like to perch on the spider vanes, and you can guess what might happen after that.

The trick here is to tilt the tube slightly which discourages birds from landing on the vanes while also generally keeping things like leaves from falling into the tube.

You should also never leave the telescope unattended while outside because a little rain sprinkle (from a cloud, a sprinkler system, whatever) can really make a mess out of the inside of a telescope.

Navigating the night sky

Whether or not you have a telescope mount that can automatically find things for you in the night sky, it can be helpful if you know your way around the night sky at least to a small degree. Why? For starters, your telescope's computer software will probably only contain a small number of objects, so some things you may want to image may not be there. A great example might be a new comet or supernova you want a picture of.

Comet C/2009 P1 Gerradd.

Another use is when you have partial clouds or a lot of light pollution coming from one area of the sky, you can easily identify coordinates where you can and cannot shoot targets.

Of course if you have a fully manual telescope like a basic model Dobsonian, celestial navigation basics become very important indeed if you have any hope of finding a specific target.

Fortunately, celestial navigation is straightforward and should not take you much time at all to get the hang of.

Celestial coordinate system

In midrange astronomy and above, most of us have completely computer controlled telescopes with high end planetarium programs that can find and track hundreds of thousands of targets including comets without even thinking about it. Unfortunately, if we are on a budget our software may not include all the targets we want to see, and even if it does, it probably will not point the telescope right to it. So how do we find our targets?

We start by understanding how to navigate the sky manually. There are two types of navigation, altitude & azimuth and right ascension & declination.

Altitude & azimuth is a very simple system that measures the angle in degrees of an object above the horizon (altitude) and the angle in degrees of an object from north in a clockwise direction.

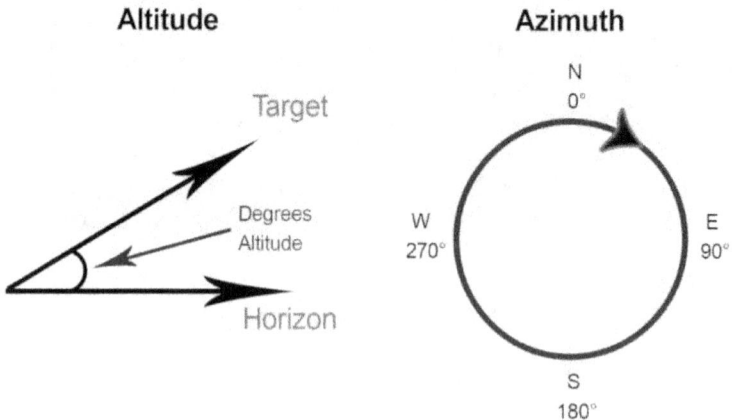

Illustration of altitude and azimuth.

An example would be the North Star, Polaris. At my location Polaris would have an altitude of approximately 30.5 degrees and an azimuth of approximately 0 degrees since it is almost exactly north.

Choosing and Using a Dobsonian Telescope

Altitude and azimuth make it very fast and easy to find any object in the sky, but it does have one big drawback and that is that these coordinates are always valid for a given date, time and location since the earth is constantly rotating (with the exception of Polaris of course). This means that if an object is at an altitude of 30 degrees with an azimuth of 94 degrees, in an hour that will have changed to different coordinates as the object will have changed position in the sky.

Think of it this way, early in the morning the sun rises in the east at 90 degrees (not really, but for the purposes of this example let us say it does). Further, let us say a day is exactly 8 hours long at this time of year. That means the sun moves at 180 degrees in 8 hours or 22.5 degrees per hour. So if the sun in our example was at altitude 0 at 8am, it would have an altitude of 22.5 degrees at 9am, 45 degrees at 10am, 67.5 degrees at 11am and 90 degrees at noon, directly overhead. Yes, that would be a really short day, but it should help you visualize what is happening with this coordinate system.

Next comes the right ascension and declination method (RA/DEC) which is a little harder to use, but does not depend on the date or time at all.

Declination is basically the same thing as latitude, or a measurement from the equator which is 0 degrees to the poles, positive 90 degrees to the North Pole, and negative 90 degrees to the South Pole. Think of this as being projected from the surface of the planet out into space, so an object that is in a direct line above the North Pole would be at +90 degrees declination regardless of where on earth you were standing. Declination is measured in degrees, minutes and seconds from the celestial equator.

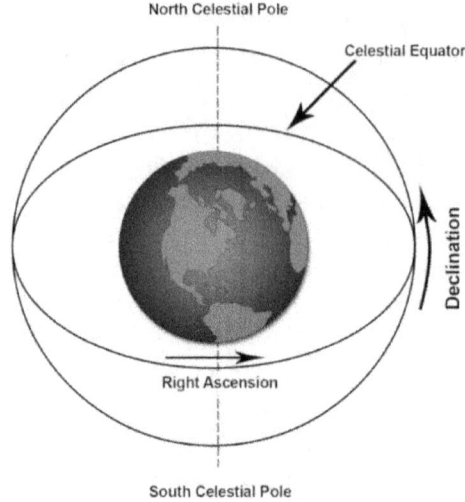

Illustration of right ascension and declination.

Right ascension can be a tough one for people to understand. There is an imaginary point in space called the Vernal Equinox which is the starting point of measurements in a circle running along the celestial equator (if you took the equator on earth and projected it into space, this would mark the celestial equator). If you started at this imaginary point and spun the earth one full revolution back to that same point that would be 24 hours of rotation, always measured at the celestial equator. Right ascension is measured in hours, minutes and seconds from that point east.

Every object has a single RA/DEC coordinate, which remains the same regardless of date or time. Messier 42, the great Orion nebula for example is RA 5h 35.4m and DEC -5 degrees 27'.

How in the world would you be able to use RA/DEC? Most equatorial mounted telescopes have both RA and Dec setting circles built into the mount. The easy method is to point the telescope at a target you know, set the setting circles for the correct values, and then move the telescope until the setting circles show you are pointed at the new target you want to find.

Seeing conditions

Now we come to seeing conditions. Have you ever noticed that some nights you seem to be able to see more stars than other nights? Some nights the stars really seem to twinkle and other times they are just still pinpoints of light? This is probably due to the seeing conditions.

Many things affect the seeing conditions such as the amount of water vapor in the air, not just down here where we are, but at one hundred thousand feet, and everywhere in between. Even the temperature at different altitudes can affect the seeing. This is why most large telescopes are built at the top of mountains or in the desert.

Now you don't have to travel to get good conditions, you just have to watch the weather. Among other things, astronomers use what is called a Clear Sky Chart (CSC) to see what the conditions are likely to be for the night. You can see what one looks like by visiting the following website and searching for an observing site near you:

http://cleardarksky.com/csk/

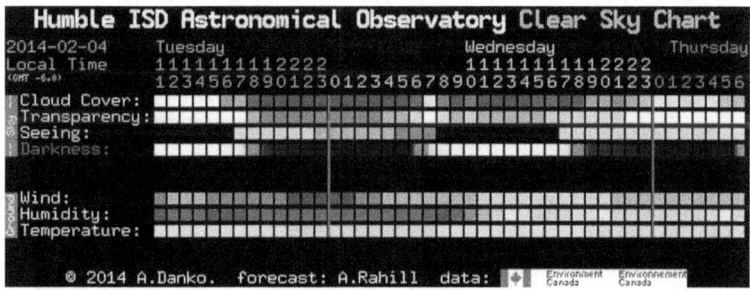

A sample clear sky chart.

The chart is read by the hour, which you can see above under the words Tuesday, Wednesday and Thursday, in 24-hour time. Each square is colored and for the top six squares from top to bottom the darker blue the better, the lighter the worse. So if you had all six of eight squares for a specific time in the darkest blue possible, that would be an estimated perfect night for astronomy or astrophotography.

So let us talk about what each box means from top to bottom. Cloud cover is the estimated amount of cloud cover for that particular time, white being completely overcast, light blue such as 16 under Tuesday in our example chart being 50% cloud cover and the darkest blue being seeing no clouds anywhere. That was the easy one. ☺

The next box down is transparency and that is pretty much a measurement of the amount of water vapor in the air. Yes, clouds can contain very high amounts of water vapor, but no clouds in the sky does not mean a lack of water vapor, or even a reduction of water vapor. Transparency is very important when your targets include galaxies and faint nebulae but not that important when it comes to clusters.

Next is Seeing and that is a measurement of the air turbulence caused by different thermal layers in the atmosphere. You may have noticed that as you look down a long stretch of road in the summer you can see that things in the distance are distorted by the heat rising from the road. This demonstrates how light can be refracted by the different thermal layers which can make it very difficult to image planets and stars. What happens is the rippling effect of looking though the thermal layers tends to make the stars twinkle, and this in turn makes them look far bigger, more bloated in an image and smears details.

Darkness is the next item down and it is just that, a box that tells you how dark it will be relative to daytime. There are basically two things that affect this, the sun and the moon. If neither the sun nor moon is anywhere near rising, this will be very dark blue. If the sun has long since set but the moon is up, then it will be lighter blue. If the sun is up, it will be white.

After darkness we have wind. The more wind, the harder it is to get a good image. Larger telescopes also have more of a problem than smaller telescopes with wind. This really needs to be dark blue to get anything accomplished imaging as anything lighter than medium blue will be 12mph or higher winds and that will probably blow your scope around like a kite.

The next two, humidity and temperature are not as important really as there is not much you can do about either. Temperature will be directly related to where in the world you are and the time of the year so normally watching it from one night to the next will make no difference. Humidity doesn't matter much as that is a measurement of water vapor at ground level and all we care about is total water vapor level shown in transparency, we don't really care whether it is at ground level or fifty thousand feet.

Please keep in mind that these are forecasts so take them with a grain of salt and be prepared for the unexpected. I always tend to carry more cold weather gear than I expect to need and also some full size trash bags to throw over things should rain come from nowhere.

There are applications that can put the CSC on your Windows desktop, apps (iOS and Android) and programs (Weather Ninja by CCDWare) to display the Clear Sky Chart along with, of course, its web page.

Finding targets

So let's assume that we have all the equipment we need to start our journey in astronomy, and more than that, we understand how to use it all. Where do we point this thing?

There was a time when you would purchase items like the *Sky Atlas 2000* by Wil Tirion for maps and *Burnham's Celestial Handbook* by Robert Burnham Jr. for the details on the objects, sit down for a few hours and plan on what target would be best to see on a specific night. While I still own both of those (they are still fun to use), those times have passed and virtually everyone these days uses computer software to plan their sessions.

If you want to start out with something physical instead of a computer you have a lot to choose from. I suggest starting out with a Planisphere such as the Millers large size available from most astronomy supply houses and of course Amazon. You need to get the model made for the closest latitude you can.

I also recommend the Orion DeepMap 600 available from www.telescope.com or Amazon.

My last suggestion is Sky & Telescope's Field Map of the Moon, which is available in both standard edition (for Dobsonians and other reflectors) and mirror image versions (for refractors and SCTs).

I carry all three of these items all the time and use them often to show new people where things are and how to find them.

Whether you prefer to use a computer or tablet, there are choices for you. I personally find that computers are generally better for the dirty work of planning sessions and tablets are better for use in the field for finding targets in the sky, looking up alternate targets, or identifying things you see. One of the beautiful things about this hobby is that you may completely disagree and find a way to do things that works better for you.

Lists of targets are generally contained in what are called catalogs. There are tons of catalogs out there, some of which may only contain one hundred or so targets, some which contain millions.

Generally speaking most astronomers, and especially ones on a budget in the Northern Hemisphere, start off with the Messier catalog which contains 110 objects. This is so prevalent that I even wrote a book about imaging them called the *Messier Astrophotography Reference*. The book is also useful for visual astronomy because it can help you find and visually identify objects in the sky.

From there you might go to the Caldwell catalog (109 objects), the Hershel catalog (2514 objects), or maybe the more succinct Hershel 400 (a selection of the 400 'best' objects from the full Hershel catalog), and more.

Even the most basic astronomy software will contain the Messier catalog, and most will contain many others. Let's take a look at a few.

Stellarium

One of the more popular software packages for finding targets is Stellarium. Stellarium is a great program to get your feet wet. Since it is a free download, very pretty and fairly small it is a great place to start. It has some telescope control if your telescope has that capability and several useful plugins such as the Oculars plugin that shows you what certain targets will look like with a given eyepiece, camera, barlow, etc. Unfortunately the target selection seems rather limited and images of targets is even more limited.

Stellarium is available from www.stellarium.org

Stellarium main screen.

Let's take a closer look at Stellarium. The images I show are windowed view, but by default it comes up in full screen mode.

The first thing we need to do after installing Stellarium is to tell it where we are. You can do this by moving the mouse over to the left of the screen and down just a little below center until the side menu slides in from the left side, which is shown in the next image.

82

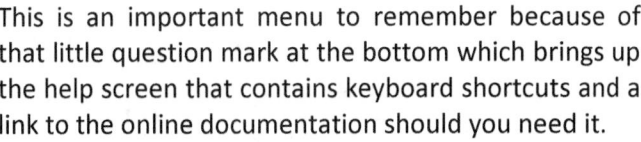

This is an important menu to remember because of that little question mark at the bottom which brings up the help screen that contains keyboard shortcuts and a link to the online documentation should you need it.

We want to click on the very top icon, the one that looks like a star. This will bring up the configuration window for our location.

From here you can search for nearby cities, click on the world map, or directly enter your latitude and longitude. When you finish be sure to check the checkbox to save this as your default location. Once you are done here you can close this window.

Now go back to the menu on the left side as we previously saw. Click on the icon of the wrench, the fifth icon down. This brings up the general configuration window. On the top of this window you have six icons. You want to click on the one on the far right named Plugins.

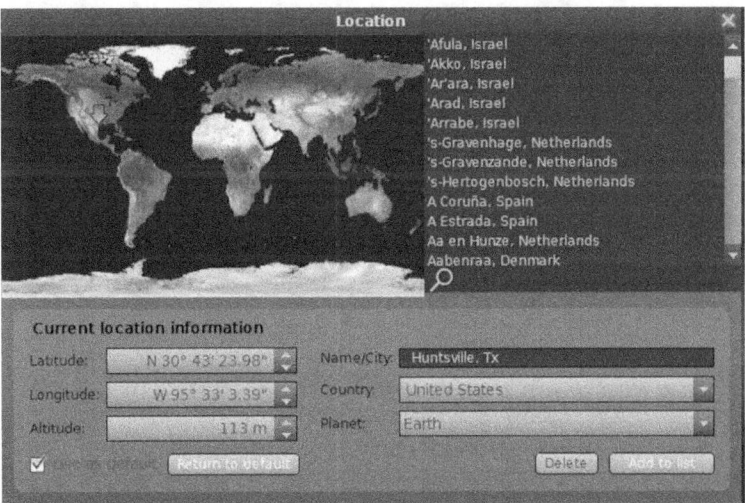

Over on the left you will see a list of plugins. Click on Oculars. Now in the bottom right corner of that screen you will see a button that you should click named configure. At the top of the

Oculars configuration you can click on Eyepieces. You should see something similar to this:

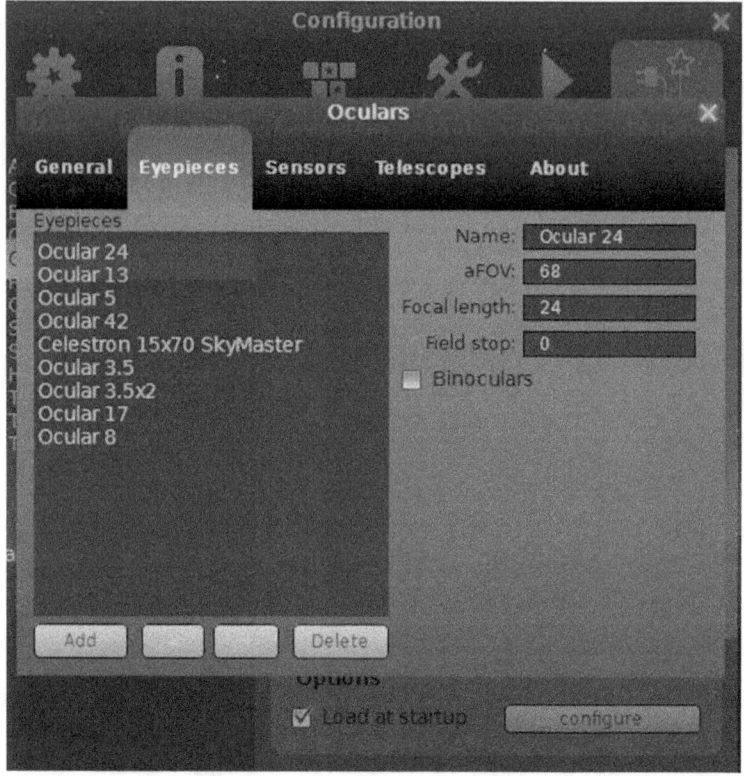

Stellarium oculars configuration.

There may be more configured in my screenshots than you have. That's fine, go ahead and click on Add and start putting in your eyepieces. Once you have some in, click on sensors and put in your camera, then finally click on Telescopes and enter the information for your telescope.

Once you are done with the configuration close all the configuration screens and go back to the main interface.

You use the bottom ribbon for general controls which you access by moving the mouse to the lower left side of the bottom of the screen. The menu will then slide up:

Stellarium bottom menu.

From here you can do a variety of things including turning on and off constellation boundaries, turning on and off both equatorial and altitude/azimuth grids, turning on and off planets/nebulas, pausing the rotation, etc.

The big feature here is the Oculars view. Move the mouse to the left side of the screen to get the left menu out, then click on the magnifying glass icon, fourth one down. You should then see the find box:

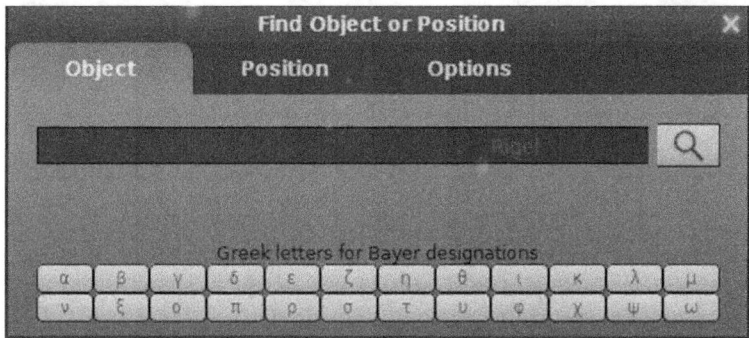

Stellarium find window.

Now we need to type in an object we know is in the sky right now. You can try M43. If it is not in the sky, choose another Messier object and continue. Let's assume it is. Type in M43 into the box and press the enter key. This centers our view on M43 and puts some relevant information in the upper left of the screen.

Now move the mouse to the bottom of the screen to pull up the bottom menu and click on the Oculars view icon, that seventh one from the right and looks like a circle in a square. That should bring up something similar to this:

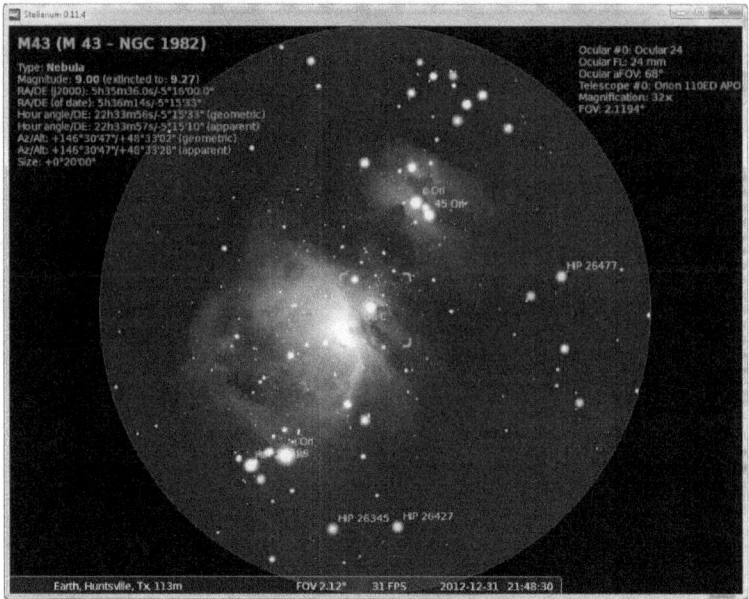

Stellarium showing M43.

So why did I say something similar, and why might your view look substantially different than mine? The answer is that you probably are using a different telescope, different camera, or a different eyepiece.

To switch between the different things you configured you press Alt-O which brings up a window like this one:

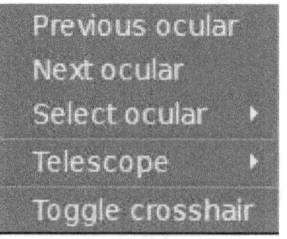

Oculars selection pop-up.

You can press CTRL and Q together to exit the program. This concludes our very short walkthrough of Stellarium.

C2A

Next up is a package called C2A that stands for Computer Aided Astronomy and is a very capable program. Even better, it is completely free. If I wanted to get by on the cheap, this would be my choice. You can download it from:

http://www.astrosurf.com/c2a/english/

While not as polished and user friendly as Stellarium, it is far more replete with features containing many object catalogs such as Messier, NGC, IC and PGC.

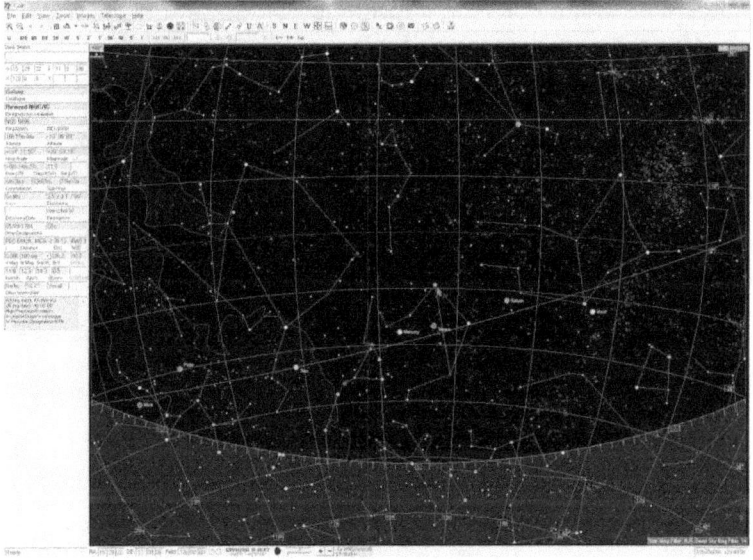

C2A main window.

Once you install C2A the first thing you need to do is set your location by clicking on the Tools menu and then Options. Once that dialog box is up click on the Location tab and set your longitude, latitude, altitude and UTC Offset. This information is required so the program can show you what is in the sky above you at the correct time.

C2A has a toolbar where you can access the most used function as shown in the next image.

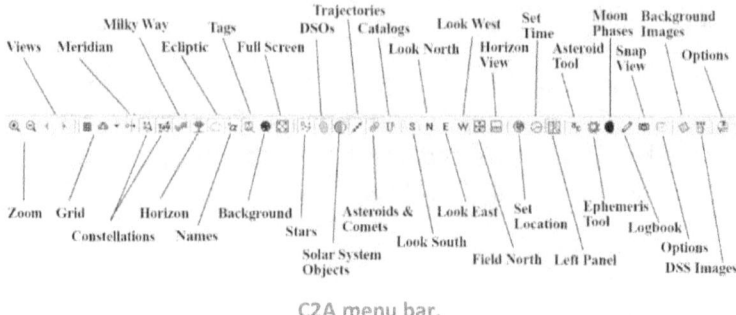

C2A menu bar.

At the top on the far left under the toolbar is a search box where you can type in a target, for example, M43, and press enter. This will fill out the information below and show you on the star chart where that object is.

C2A object information panel.

The object you searched for will also now be centered on the screen. Deep sky objects, such as the M43 we just searched for, will show up with outlines and if you have the option turned on (Tools->Options->Deep Sky->Labels) then the outlined objects will be labeled as in the following image. The outlines are color coded: green for nebulae, red for galaxies, pink for clusters and possibly others I have not seen yet.

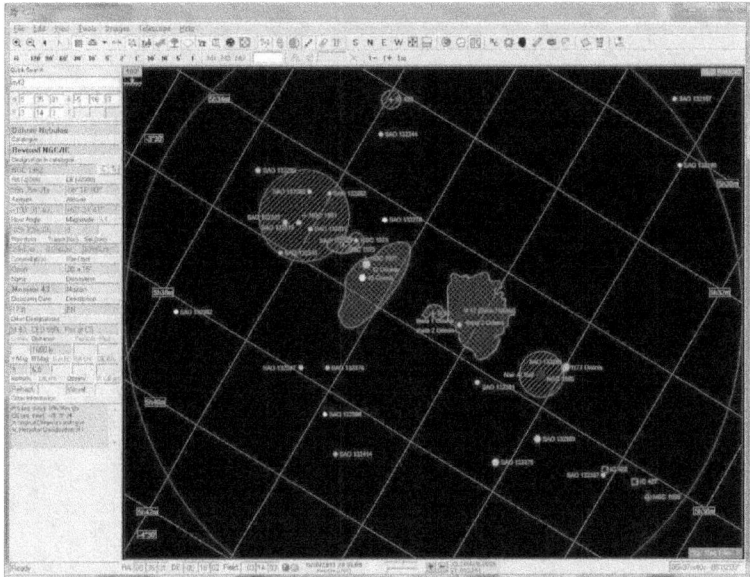

C2A showing M43.

Navigating the star chart is pretty easy; you can use the mouse scroll wheel to zoom in and out, right click and select center to center the screen on that location. There are four buttons in the center of the tool bar labeled S, N, E and W, and these point you towards the horizon in that direction.

If you want to scroll around the sky you can use the arrow keys on the keyboard to move up, down, left and right as you would expect. The CTRL and ALT keys are modifiers in that if you hold down the CTRL key while pressing an arrow key, you will move in smaller increments than pressing the arrow key alone. The ALT key is the opposite, allowing larger movements.

To get information about an object on the screen, simply double click it and the boxes on the left will populate with the information on that object.

If you right click on an object and select Visibility you will see a chart similar to the following appear:

89

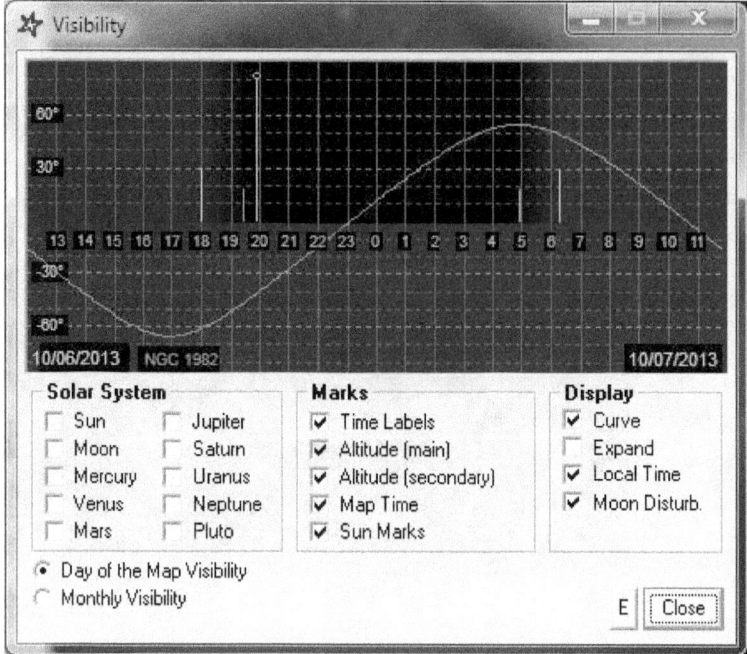

C2A visibility chart.

This chart shows you when the object will be visible in the sky at your location. In this example it shows the object rising at about 22:30 local time reaching zenith at around 05:00 just before sunrise at which time it will be about 55 degrees high.

If you have a computerized telescope and carry a computer with you into the field, you could also use C2A to control your telescope. Once you have C2A talking to your telescope you can find targets in C2A, right click on them and then select Telescope and then Slew To Object as shown in the next image.

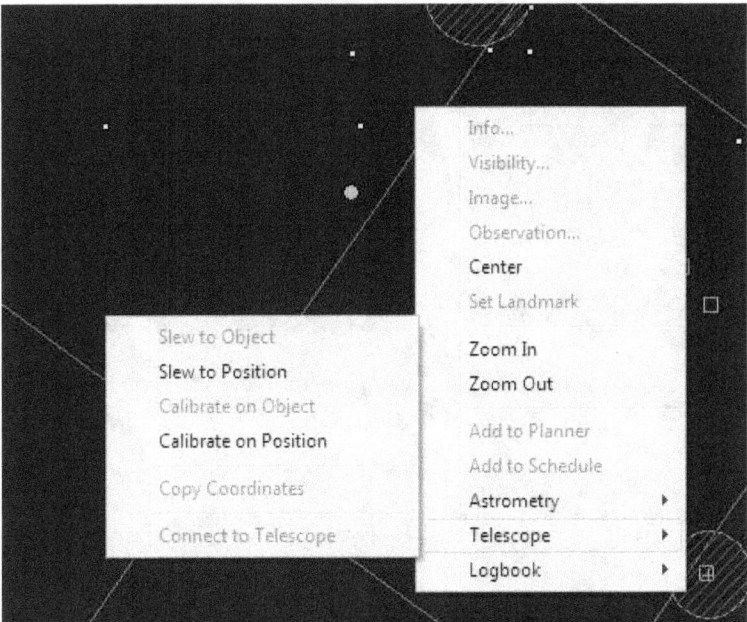

C2A right click pop-up menu.

This could of course require additional cables and installation of drivers for your telescope which is far beyond the scope of this book.

One feature that comes in very handy is the moon phases function. This brings up a chart that shows you the phases on the moon on the days of the month, complete with the percentage of full moon.

I tend to use the moon phases display when planning sessions to shoot so I have an idea of the best nights to schedule.

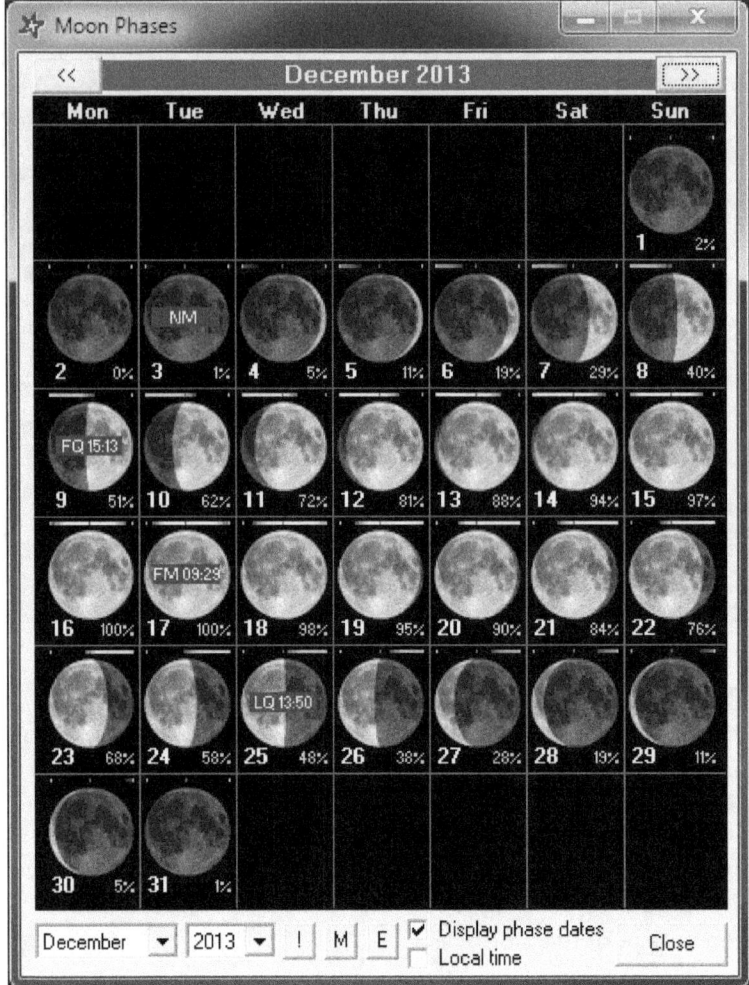

C2A moon phases window.

As you can see in the above image it displays the moon phases on a calendar. You can use the arrow buttons at the top left and right to move through the months, or use the drop down boxes at the bottom left to select the month and year you want directly.

Three buttons are at the bottom center and they are: "I" sets the calendar to the current month and highlights the current date, "M" does the same thing except uses the date set in the program's star map, and "E" exports all the days as individual little images although I am not sure why.

One of my favorite features of C2A is accessed by clicking on Telescope and then Observation Planner. I have no idea why this feature is under the Telescope menu, but it is.

C2A's observation planner is fast, easy, and very flexible. Choose an area of the sky before launching it (or you can check the box at the top right of the observation planner that says "No search limit" to use the entire sky) as it will use that part of the sky to compile its list. Then select what you want to view in the upper left section, then any criteria such as altitude and azimuth limits, magnitude limits, size limits, and if you selected a star catalog, spectral type on the far right center, then click the red checkmark to see the list.

Once you are done, you can even print the list out to take with you in the field. This makes a handy list of things you might want to image that evening.

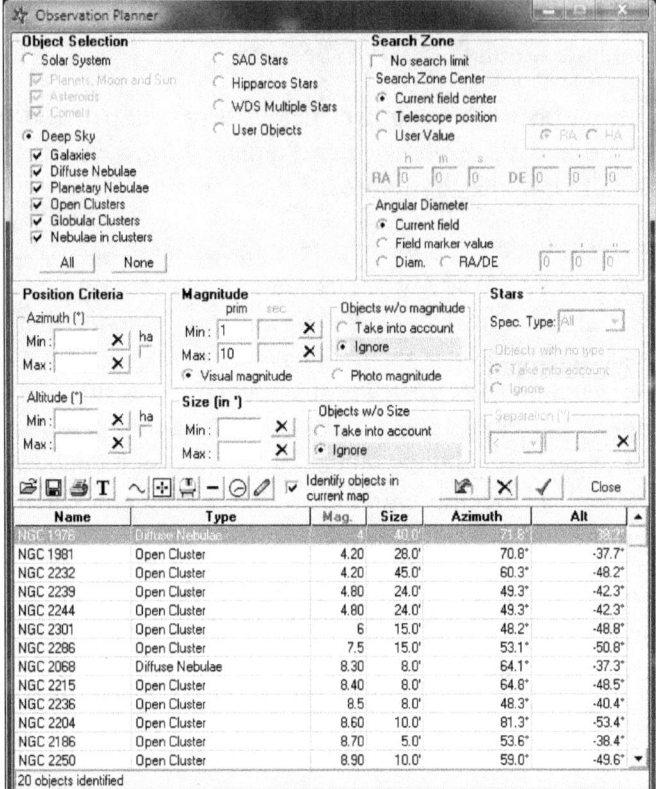

C2A observation planner.

Another nice feature is the ability to see DSS (digitized sky survey) images of a target so you can get an idea of what that target looks like (this requires an active internet connection). To make use of this simply do a search for an object (in this example I did M17). Once it is centered in your field of view click Images->Digitized Sky Survey. That will present you with this dialog box:

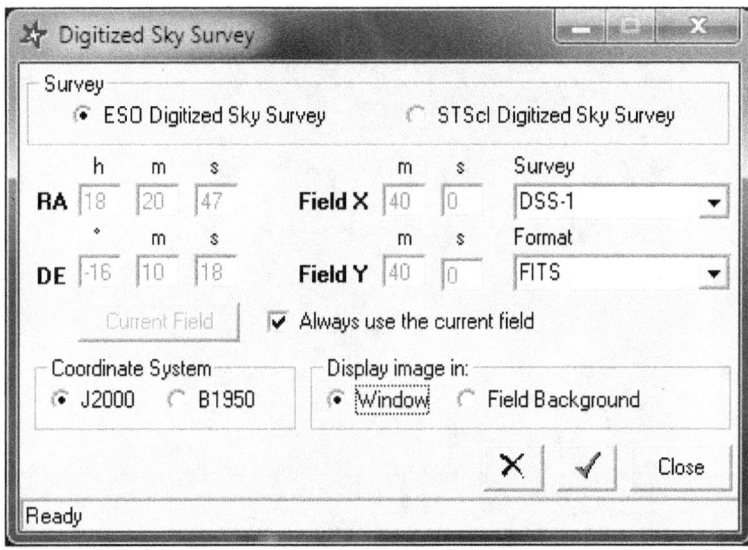

C2A DSS display configuration window.

Make sure that in the lower right side, "Display image in:" has "Window" selected and click the check mark. This will download and display the DSS image for this target assuming one is available.

C2A DSS display window.

Since we are imaging, a very useful tool is a feature called Field Markers (Tools->Field Markers or CTRL+R) shown in the next image. This allows you to display the area on the map that your camera will capture in an image. This can be very useful for figuring out if a target will fit completely on your camera's sensor, or if the target will be too small to really see anything.

You will need some information for the configuration including the focal length of your telescope in meters (my primary imaging scope has a focal length of 770mm or .77 meters), the physical size of your camera sensor in mm (most camera manuals will have this information in the specifications section) and the size of the photosites (the tiny pixels that capture the light, also usually in the camera specifications section of the manual).

Once you have the pertinent information, plug it into the following dialog box under the CCD Field Rectangle section and make sure the checkbox for Rectangle 1 is checked as shown here.

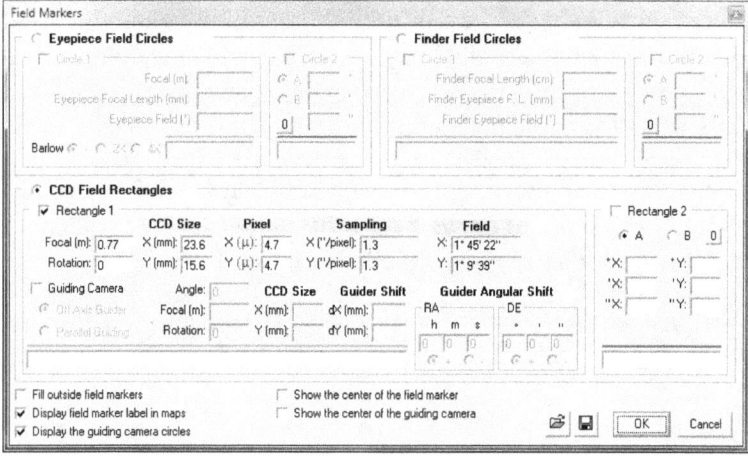

C2A field marker configuration window.

While you are here, notice there are areas for eyepieces, finders, and even a section for an arbitrary rectangle where you set the size. Once the dialog box is complete, click OK and you should see a rectangle such as this:

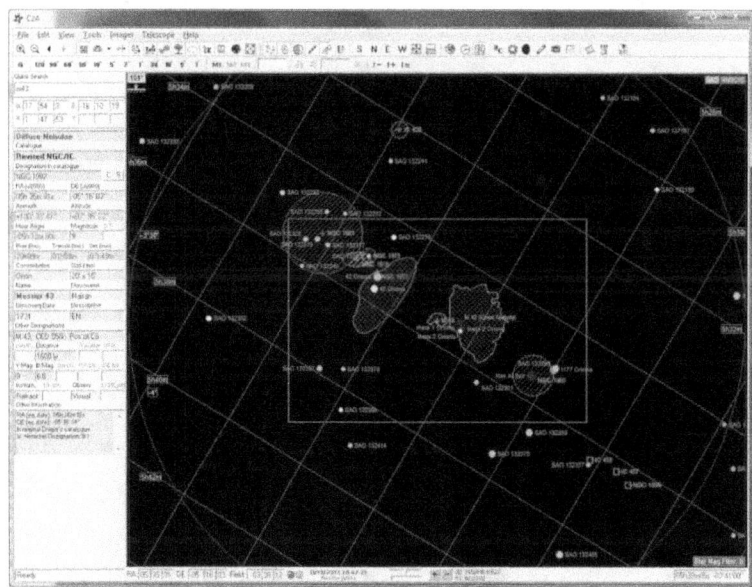

C2A displaying M43 with field markers.

In this case you can see that my camera would capture M42, M43 and a couple of other targets in one frame.

Choosing and Using a Dobsonian Telescope

Since pretty much all I do is astrophotography I generally keep this feature enabled all the time, but if you are doing visual astronomy you might want to disable it or switch to an eyepiece view using the same dialog box we were just in.

There are a ton more features and capabilities in C2A but this should be enough to get you started. I would suggest you check out the author's tutorials at the following address:

http://www.astrosurf.com/c2a/english/support.htm#Tutorials

Tablet software

Tablets and smartphones, specifically iPads/iPhones, are so prevalent in today's world that I just could not leave them out. Too many times I have enjoyed both visual and AP with the help of my iPad and iPhone. From navigating the night sky, to seeing the moons of Saturn orbit the planet in a planetarium, to just watching videos while my scope imaged, the tablet and smartphone have become a staple when I go out.

Why? One thing a laptop cannot do well is be lifted up and compared to the sky and then move in real time as you move it against the sky. Tablets and smartphones excel at this. Until you have tried this you have no idea how cool it is to put a moving sky chart up to the sky and move it around to see what all is in that area of the sky. This is absolutely wonderful for visually exploring with binoculars or a telescope, or it can help you explore targets that are in the same area of the sky as targets you are already shooting.

Star Walk main screen.

Like desktop planetarium/star charting software there are several options for this function on tablets. To start with there

99

are programs such as Star Walk ($4.99 iPad version) which is a planetarium and planning guide which lists all of the Messier objects and some others. The database is small but the graphics are rich, and for the money, this is my choice of apps for the casual user. It also has a wonderful feature called Sky Live which shows you the rise/set times of the Sun, Moon, Venus, Mars, Jupiter, and Saturn at a glance as well as their angle in the sky, and phases of the moon. Here is the Sky Live screen:

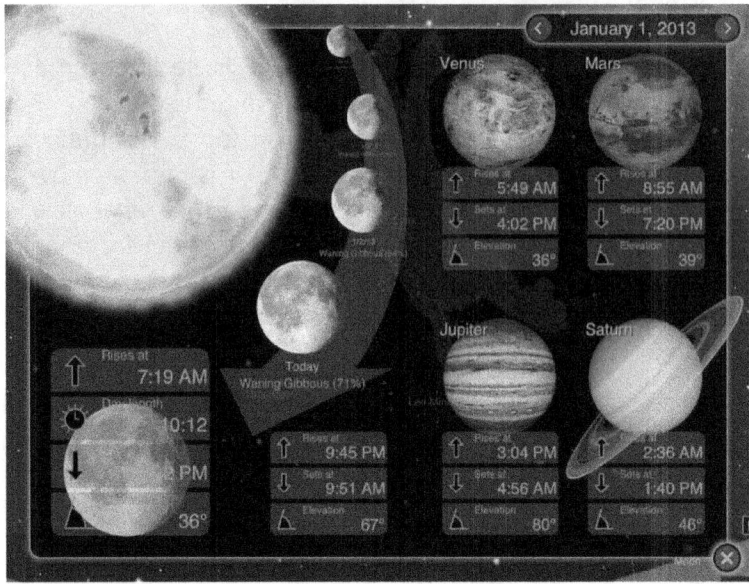

Star Walk's Sky Live screen.

More information on Star Walk can be found at their website:

vitotechnology.com

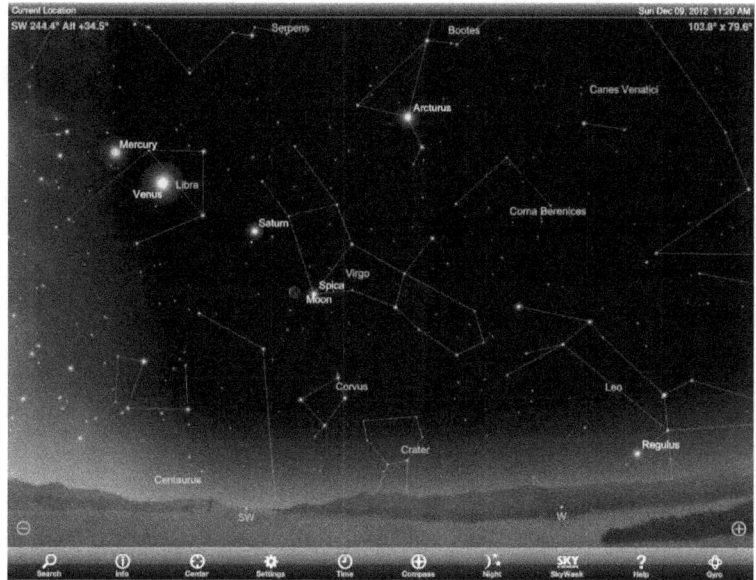

SkySafari's main screen.

Next up on the list is SkySafari ($2.99, iPad/iPhone/Android). This is the entry level version from Southern Stars and is a very capable program. While not as "pretty" as Star Walk it is very easy to get down to business with, and with over twice as many objects in its catalog as Star Walk, it means business.

One step up is SkySafari Plus ($14.99, iPad/iPhone/Android). This app lists 2.5 million stars, telescope control functions, and a list of 31,000 space objects of special interest! It is just as fast and easy as the standard SkySafari, but contains tons more information.

From there we can jump to SkySafari Pro ($39.99, iPad/iPhone/Android) which boasts one of the largest stellar databases of any planetarium program for any platform (including PC/Mac!) .

My favorite feature of SkySafari (other than its massive object database) is the search feature. Oddly enough, this feature is not for searching:

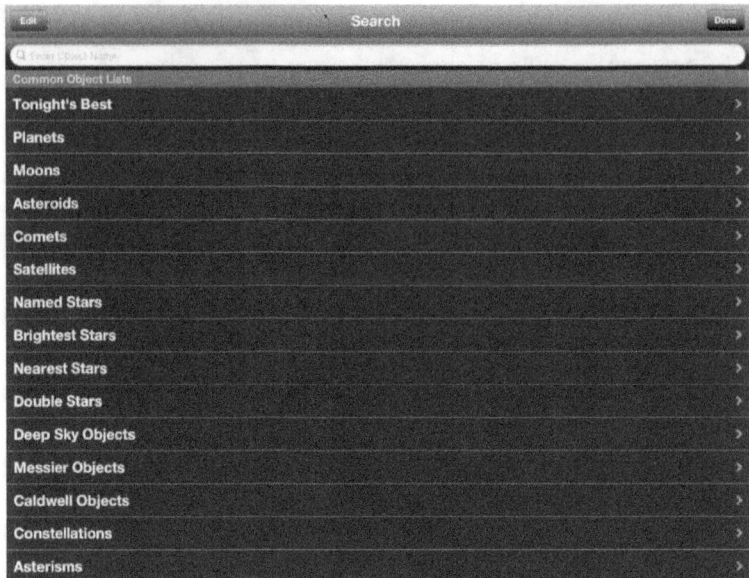

SkySafari's search screen.

Once you click on the little magnifying glass in the lower left of the SkySafari main screen you are presented with this screen. Here of course you can start typing what you want to search for in the search box at the top. But that isn't the cool part.

Notice the items listed below the search bar. Here you can tap on Satellites, for example, and you will see a massive list of satellites with the ones that are currently visible highlighted. You can do the same for Asteroids, Planets, Comets, whatever. Of course you can tap on Tonight's Best for a listing of the best objects to view in the sky as well. I love this feature!

For more information on SkySafari and other products by Southern Stars, see their website:

www.southernstars.com

If you are less interested in having a star catalog but would like to explore our solar system, watch space related videos and keep up on the latest in the space program, the NASA HD app is a wonderful little app, and it's free! It also has a detailed satellite tracker.

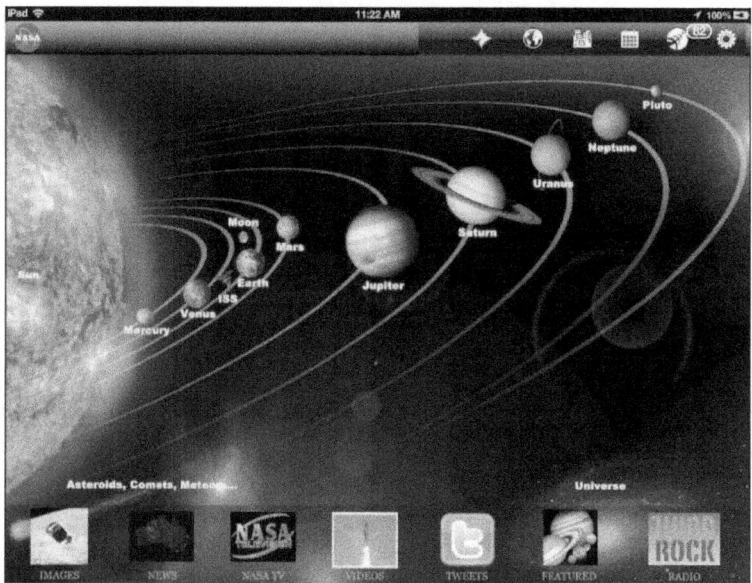

NASA HD main screen.

More information on the NASA HD app can be found on their website:

www.nasa.gov/centers/ames/iphone/

Other useful apps are things like ICSC Clear Sky Chart to help you know when you can go out imaging, Moon Globe for finding features on the moon, and GoSkyWatch planetarium, all free.

Another nifty app is AstroAid by Paul Rodman, the same author as AstroPlanner discussed elsewhere in this book.

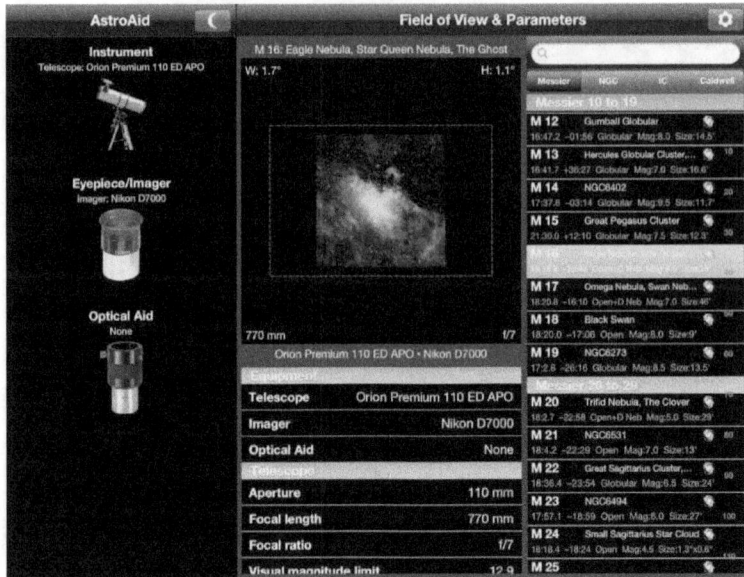

AstroAid parameters screen.

This little app will let you put in your scope/lens information, your eyepiece/imager information, and then get a realistic approximation of the view for that configuration. That can be very handy.

You can get more information on AstroAid by visiting its website at:

http://www.ilanga.com/astroaid/

Power sources

Another very important question is how will you provide power to your accessories, assuming they need power? I am very fortunate in that I observe from an observatory which has A/C power so all my equipment is plugged directly into A/C with the exception of my DSLRs which run on their own batteries, and I carry two fully charged batteries for each camera when I image.

If, however, I go chasing transits and eclipses like I did in 2012 I will need portable power and for that I use large battery packs.

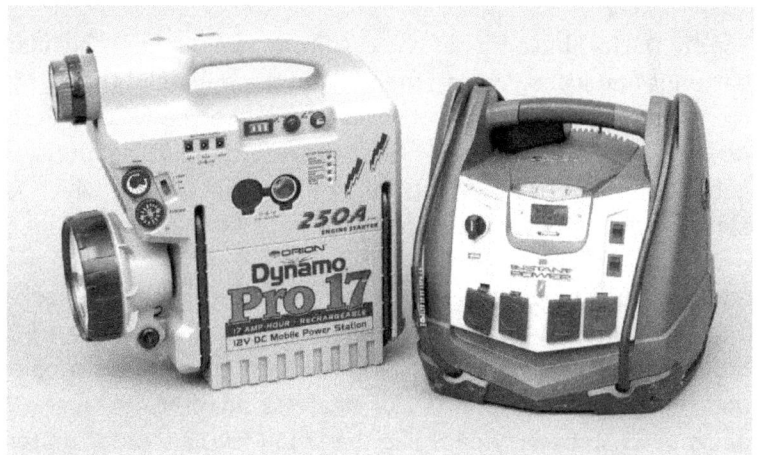

Battery packs, Orion Pro17 on the left, Schumacher XP2260 on the right.

One is a 17Ah (amp hour) Orion battery pack and the other is a Schumacher Electric XP2260 22Ah I purchased at Wal-Mart. Both cost about the same. The Orion pack was purchased with my scope and was bought for the warranty, and because should anything go horribly wrong it would be very difficult for tech support to blame the power to the scope for anything that happened ☺

Battery packs like these are measured in amp hours, or the amount of power a battery pack can provide over a given amount of time. For example, a 17ah pack can supply 17 amps of power for one hour, or 8.5amps for two hours. Although the math works out like that, in reality as you stretch the time out, the power drops a little so that you do not get the full amount.

Something you should be aware of with these types of battery packs is they need to be kept charged. If you let them drain too far down they may not recharge with their normal chargers. What I personally have had to do is disassemble the unit, remove the battery, and charge it on a high power battery charging unit I own for cars, boats and small engine machines like lawn mowers. It does 12v and 6v using a variety of charging options. Use this method at your own risk! This will also void your warranty.

For my DSLRs I have lots of choices, from using multiple regular batteries and just swapping them out in the middle of the night, to running off AC power with something like the Nikon EP-5B power supply which replaces the battery with an AC adapter, or getting an extended runtime Nikon EN-EL15A battery which is almost twice the size as a standard battery for that camera.

So how much power will you need? How large a battery pack? That isn't an easy question to answer. Let's start with an EQ mount and go from there. If you have an Orion Sirius mount and run it off the supplied 12v cigarette adapter, that uses 2 amps. A 17Ah battery pack like the Orion Pro17 should run for about (17/2=8.5) eight hours depending on the temperature and if the pack has a full charge, and how much you slew, etc.

I use an AC to DC inverter for my laptop which when plugged into a wall uses 2.5A of power. You lose quite a bit in conversion so figure 5A of use. If that were the only thing plugged into the 17Ah pack it should last three hours.

Dew strips can pull quite a bit and I use four which pulls about 2A. Again, if they were the only things I run, then about eight hours.

Now we add all this up, 2A+5A+2A = 9A, less than two hours on the 17Ah pack on a good day with warm temperatures. Realistically? About an hour.

One hidden thing, or at least something many people forget about is that their guide camera probably draws power from USB which means it draws laptop power, which drains your laptop battery much faster than if you were just sitting at your desk. Be prepared for that.

If I really wanted to shoot all night off batteries I would have three battery packs, all 22Ah, or at least the 17Ah and 22Ah I have plus another one of the 22Ah using the two large ones for dew control and the laptop, the smaller one for the scope.

One word of warning here, many battery units have a hard time supplying more than 2A to an outlet, be it the cigarette lighter type or the AC outlet type. For example when using my larger laptop (2.5A AC) even the larger 22Ah pack has a hard time running it for very long before it starts screaming, probably a thermal warning. The only time I am really using it in the field on battery at the moment is doing solar (the eclipse and transit of 2012) so both times I could run the car, use a DC to AC inverter in the car, run an extension cord over to where I was and plug the laptop in there.

You could of course also use a generator but I would not advise showing up to a star party with one, your noisy motor will not be welcome. If you are going to use a generator you may want to consider something like the small Honda generators which are very quiet, (for a gas engine that is).

Odds are, when you are starting out you have none of the equipment I just mentioned. That's OK, as you get more equipment you will know exactly how to find out how much power in a battery you will need over the course of an evening.

Astrophotography

While the Dob is not a good tool for astrophotography, that does not mean you cannot get some good pictures with one. What it does mean, is that you need to be realistic with your expectations.

The following sections will not only get you started, but will probably provide way too much information for you right now. Hopefully this will allow you to reference back here as you get more advanced.

To start with, there are two basic types of astrophotography you can do with a still camera (pictures, not video); short exposure and long exposure.

Short exposure is when you have your shutter open for 30 seconds or less (assuming you can control your shutter that way). Keep in mind that when I talk about opening a shutter, that same theory applies to cameras that do not have a shutter but instead just activate a sensor.

Short exposure astrophotography is suitable for pictures of the moon and a few deep space objects such as the Orion Nebula and Andromeda Galaxy.

Long exposure astrophotography is any exposures of 30 seconds or more, and can go for ten, twenty, thirty or even more minutes. These are suitable for the vast majority of deep space objects out there.

A Dob can successfully be used for short exposure astrophotography without substantial extra equipment or costs. They are not suitable for long exposure astrophotography although they can be forced to do some with enough time, effort, and money.

Long exposure astrophotography with a Dob is much like walking a hundred miles on a gravel road with bare feet. Can you do it? Sure? Why would anyone in his or her right mind do it? I have no idea.

So why exactly is a Dob a bad choice?

The first reason is that the best astrophotography with a standard camera is done with a DSLR in what is called prime focus. Prime focus is simply where the lens for the camera is removed and then the telescope is attached as if it were a lens.

The biggest problem here is backfocus, or lack thereof.

In the above images the lines represent the base of the focuser (the left line) and the point at which an image seen through the scope becomes in focus (the right line).

See how much closer the lines in the left image are than the image on the right. This means in order to get the image in focus in a camera, that camera must be much closer to the telescope tube than an eyepiece would be.

Virtually all Dobs are not designed for the focuser to travel that far in and thus, the telescope cannot focus with a camera attached. This is called a lack of backfocus.

There are four ways to fix this issue. First, insert a barlow between the camera and the focuser. This will increase the magnification, reduce the image quality a bit, but will increase the focal point so that you can achieve focus.

The second method is to get a low profile focuser. This is a focuser that is specifically designed to get as close as possible to the telescope tube which in most cases will allow a camera to reach focus.

Third is you can remove the primary mirror assembly, cut an inch or so off the end of the telescope tube, drill some new attachment holes and remount the primary mirror. I highly advise you not to do this. The odds are very high that you will wind up making a mistake and the telescope will never operate properly again.

The fourth method is using an eyepiece projection adapter and an eyepiece. We will cover this in more detail shortly. The down side here is that it adds some distortion to the image. It is basically a method of afocal astrophotography specifically for DSLRs.

Another reason Dobs are not good for astrophotography is that they were never designed to have the additional weight of a camera mounted on them. You can solve this by adding weights to the base to help offset the camera weight but this is not as easy as it may sound. You can also buy a telescope such as the Zhumell which has excellent tensioners that will dramatically help instead of the cheaper tensioning mechanisms used by Orion and others.

Lastly, they are huge buckets which are not very stable. The slightest bump, wind gust, vibrations from people walking by, anything really, makes them shake and shimmy. This is horrible when you are trying to take pictures through a telescope.

Fortunately there are other methods you can use which we are about to cover which do not involve all this mess and can still provide you with some images you can play with.

Afocal astrophotography

Afocal astrophotography is basically just holding a camera up to the eyepiece of a telescope and snapping a photo. It sounds simple enough but the reality is that there are also many different types of adapters to hold your phone, tablet or virtually any type of handheld camera up to the eyepiece of a telescope.

One advantage to the afocal method is that it can solve an issue where a telescope that was not built for astrophotography cannot achieve focus using the prime focus method, can achieve focus using this method.

Some disadvantages include image degradation due to the increased obstructions in the optical path (no glass is perfectly clear, and the eyepiece was never intended for this use), decreased contrast due to increased reflections inside the light path (again, the eyepiece was never intended for this, nor was the camera) and possible inability to focus on some telescopes.

111

Phone/Tablet cameras

These days almost everyone has a cell phone that can take pretty good images, can you just use that? Of course you can! There are some really neat mounts for it such as the iPhone mount by Orion shown in the next image. Originally for an earlier iPhone, a little tweaking to the retainer clip is all that is needed to use it with newer models as shown.

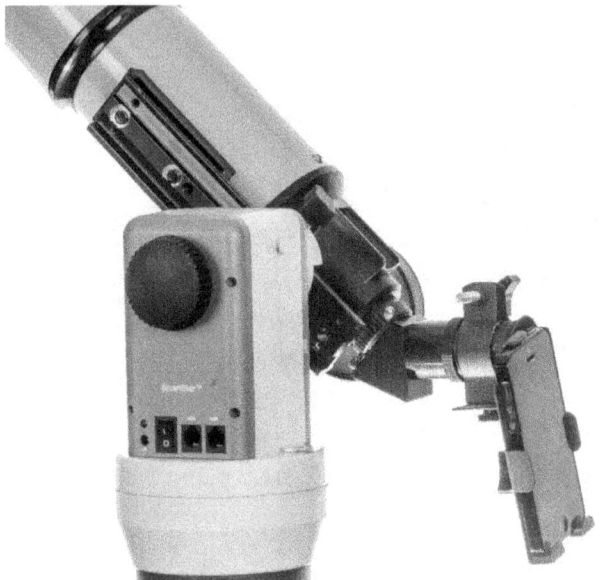

A smartphone attached afocally to a telescope.

With this adapter there are two knobs on the rear of the mount and one on the eyepiece clamp. This allows you to position the camera lens in the center of the eyepiece regardless of the eyepiece being used. I have found that if you use a light touch, you can actually tap the screen to take the picture without causing too much vibration, it works surprisingly well, although I recommend using a timer instead.

Point & Shoot cameras

Point & shoot cameras are the typical digital cameras in use today:

An example of a typical point & shoot camera.

While point & shoot cameras suffer from some of the same issues as bridge cameras, they are light weight enough to be used with inexpensive adapters pointed at the eyepiece. I have seen some remarkable images captured with little point & shoot cameras.

Many manufacturers produce different ways to mount a point & shoot to a telescope; my favorite so far is the Zhumell as shown in the next image. While a little large, it has the advantage of being so flexible I cannot imagine a small camera that could not be used with this mount. It is a few dollars more expensive than some adapters, but I would prefer to buy this thing once than buy another and have it not fit a different camera I wanted to use down the road.

A point & shoot camera attached afocally to a telescope.

If you look closely at the previous image you can see the knob at the top of the adapter which tightens a clamp onto the eyepiece (it must be a 1.25" eyepiece, 2" eyepieces will not fit) and a knob on the bottom of the camera which not only affixes the camera to the platform, but allows it to move forward and backward on the platform. Look closer at the bottom and you may see the small knob that allows the camera platform to move up and down. Not visible is the knob on the far right side that moves the platform left and right.

Make sure that your camera can disable the flash, and if it cannot, that you find a way to completely cover the flash so that no light leaks out (a word of warning; flashes can generate a lot of heat, make sure you do not trap in the heat as well as the light or you could damage your camera and/or get burned). The flash can ruin your images and really upset other people at the dark site.

DSLR (Digital Single Lens Reflex) cameras

A DSLR is a Digital Single Lens Reflex camera, one that looks like what the professionals use. You look through a viewfinder in the back (the pentaprism) and see through the lens exactly what the camera sees. The lenses are removable and can be changed out to provide for different focal lengths (focal length is what determines how large an object appears to the camera) and capabilities.

A DSLR camera.

Liveview is the ability of the camera to show on the back screen what the camera sees out the front of the lens in real time. Not only does this make it easier to see brighter objects such as the moon, Saturn and Jupiter, but it also helps us focus (more on that later).

An example of liveview is shown with the moon in the following image.

An illustration of a DSLR camera using liveview.

Full manual control refers to the camera's ability to allow you to set all the options such as shutter speed and ISO without regard to what the camera thinks it should be. For most DSLRs manual mode is achieved by placing the top rotating dial to the "M" position as shown in the next image.

The primary control dial on a DSLR camera.

In the previous image you may also note that there is a second lower dial, this is used to control the speed at which the camera takes images. The S stands for Single, or pressing the shutter release causes one image to be taken. The CL stands for Continuous Low speed which means that one press of the shutter will take an endless succession of images at low speed. CH stands for Continuous High speed. You should always use single in astrophotography.

Virtually all DSLRs also have the ability to be controlled remotely, either through software connected to a computer, through an intervalometer which is a device for taking a series of images automatically with no intervention from the user once the process is started, or by using a remote control.

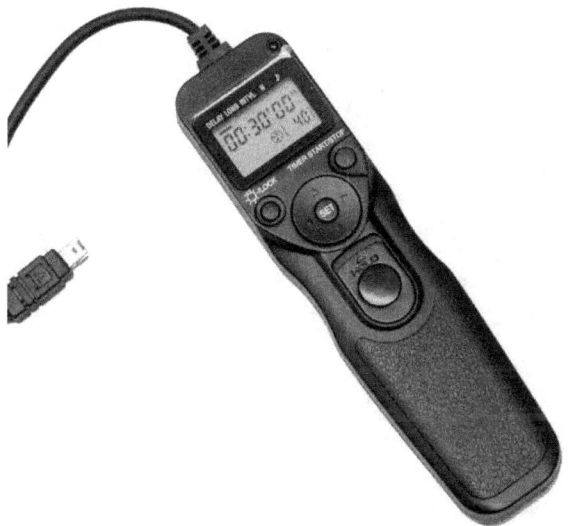

An Intervalometer.

Shown in the previous image is an intervalometer for select Nikon cameras which costs less than $40 from many online vendors. It is also available for many other makes and models of cameras, and of course you can spend more, or less, on one depending on features and quality. This one will allow me to take hundreds of images over many hours.

You can of course also fire the shutter using software if you have a computer and the correct cables. The advantage of the intervalometer is that it is cheap (compared to a computer anyway), easily portable and has low power requirement (mine has shot more than twenty hours of images on the AAA batteries that came with it), and it fits into your pocket.

A DSLR attached to a Dobsonian using a prime focus adapter.

Lastly, since DSLRs have removable lenses, they can be used in what is called "prime focus" as shown in the previous image which means a telescope can be used just as if it was a large lens. This is the preferred method because it has less distortion caused by objects in the light path (such as eyepieces which were never designed to be used to shoot images through). This method can also be used to mount a light pollution filter on the prime focus adapter assuming you purchase an adapter that is threaded for this purpose.

If you are even remotely really interested in astrophotography, even a cheap old used DSLR will open up a lot of possibilities to you. Something like a Nikon D70s (not the D70) can be had used for around $100 and will be a huge improvement over a phone, inexpensive point & shoot, or any other camera that cannot give you full manual control and be triggered with a remote. Maybe even splurge for something like the D80 which will give you live view as well for only a few dollars more.

If you want a cheap DSLR for this, personally I would spend a little more yet and get the Nikon D90 (less than $200 off eBay is common) or a Canon equivalent. The D90 is old, yes, but it is still an awesome camera that is rock solid and provides excellent images for the money.

You can also still get batteries, remotes and any accessories you could ever want for the D90.

I would not do much more than that for a Dob. If you already have a better camera then by all means, use it. Newer cameras will be more sensitive (particularly full frame ones) and provide better images at higher ISOs.

You will still be limited to 30 seconds or less with any camera you put on the Dob so don't go crazy. If however, the astrophotography bug bites you, a Nikon D610 or D7500 are awesome daytime and astrophotography cameras you will love.

Eyepiece projection

Eyepiece projection means that a telescope is fitted with a special adapter which you then place a standard viewing eyepiece inside of it, it is then attached to a camera with its lens removed. This has the effect of the eyepiece projecting the image directly onto the camera's sensor. The following image shows an eyepiece projection setup for a DSLR camera with the DSLR on the left, the eyepiece in the center and the eyepiece projection adapter on the right.

Note the silver knob on the bottom of the adapter; this screws in to tighten against the eyepiece so it does not slide around. Failing to tighten this can result in the eyepiece sliding into your camera and damaging or even destroying it.

Also note that there is a removable adapter ring on the left side of the eyepiece projection adapter. This ring is camera brand specific just like the one for prime focus astrophotography (in fact, this is exactly the same ring in both photos, I use that one ring for both types of astrophotography).

An eyepiece projection adapter and eyepiece for fitting a DSLR to a telescope.

One advantage to the eyepiece projection method is that just like the afocal method, it can solve an issue where a telescope that was not built for astrophotography cannot achieve focus using the prime focus method.

Although not as severe as with the afocal method, this method still suffers from image degradation due to the increased obstructions in the optical path (no glass is perfectly clear, and the eyepiece was never intended for this use), decreased contrast due to increased reflections inside the light path (again, the eyepiece was never intended for this, nor was the camera) and possible inability to focus on some telescopes.

One thing to note is that only certain eyepieces fit into these adapters. They are usually from 15-25mm in focal length, and most of them are called Plossl eyepieces, which are typically on the lower end of the quality scale.

This is becoming very popular with people using the more inexpensive solar dedicated telescopes such as the Personal Solar Telescope, or PST as it is more commonly known.

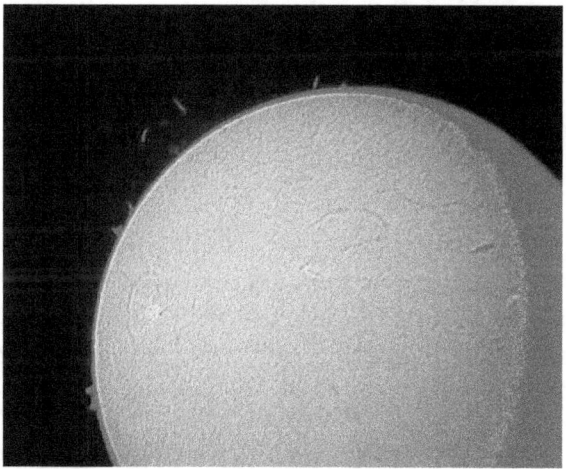

The Sun shot with a DSLR using the eyepiece projection method.

Prime focus

Prime focus is what most intermediate to advanced astrophotographers use and basically that means you remove any lens from your camera and use a telescope as the camera lens.

To attach a camera using prime focus, you normally use a prime focus adapter such as this one:

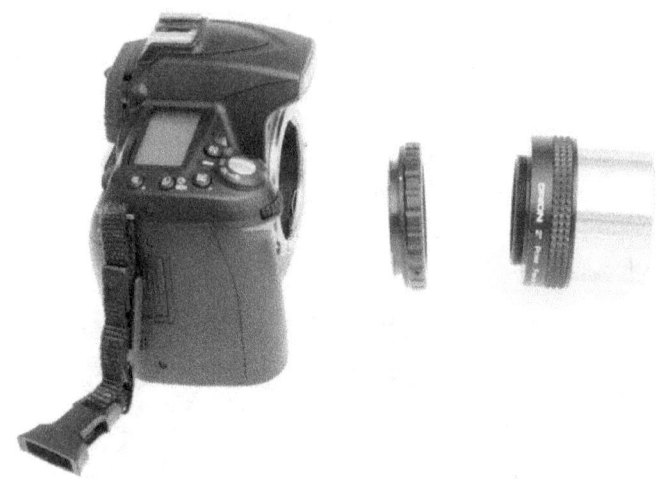

A 2" prime focus adapter to fit a DSLR onto a telescope.

This one is an adapter for a DSLR (others exist for other types of cameras) and consists of the DSLR on the left, an adapter ring that is camera brand specific, and then a fairly generic 2" prime focus adapter (they come in 1.25" as well for smaller scopes).

Prime focus has the advantage of having nothing in the optical path that isn't supposed to be there and hence the best image quality of any of the astrophotography methods. It is also the easiest to implement assuming your telescope was designed with astrophotography in mind which of course, a Dob was not.

When shopping for a prime focus adapter, get a 2" if your telescope will accept it, 1.25" if not. The 2" is more stable and will provide a better image near the edges because it will have less vignetting (fading towards the edge due to being too small a tube for the light to pass through).

Example of vignetting.

Also be sure to get one that has filter threads on it so you can use a light pollution filter (as we already discussed).

Let's clear up some issues before we get in too deep. Can you use a point & shoot (camera without a removable lens)? Sure, but your results will suffer greatly if you decide to get serious about astrophotography. All that extra glass that was never meant to focus through an eyepiece or another objective degrades the image horribly.

How about one of the newfangled mirrorless cameras (MLCs)? Sure, assuming you can find a T-Ring adapter for that camera, software that will run it and download images from it, a camera that can shoot RAW and software that can decode the RAW it shoots. Oh, and unless it is a Micro 4/3 your sensor will be so tiny that its sensitivity and noise generation will be horrible.

A great feature to have is called LiveView. This is the ability to view a video image of what the camera sees in real time, either on the screen on the back of the camera or in a program on a connected computer. This can make initial focusing much easier.

Next is the ability to shoot in RAW. Not to worry, virtually all DSLRs have this capability. A RAW file is pretty much what it sounds like, the raw data from the camera's sensor is written to a file with no adjustments, no corrections, and above all, no compression.

Jpegs (or JPGs) that your camera creates have all of those done, adjustments, corrections and compression. They can technically work, but because they are manipulated in camera and compressed they will always give far inferior results to RAW images.

One thing you notice I did not mention with cameras is their maximum ISO. Having higher ISO performance is good, but not because you want to use it. My D7000 captures outstanding daylight or nighttime images at ISO1600, and quite good images at ISO3200. ISO6400 is usable in a pinch. The problem is, you will be stretching the image so you need the dynamic range (dynamic range and more is discussed in detail a little later) offered by the lower ISOs which is why the vast majority of my imaging is at ISO800. If I need more light I just extend the exposure time.

Eventually you will hear of people "modding" their DSLRs for better response to red. Let me first dispel a myth about DSLRs. Some people say that they are completely insensitive to high wavelength red such as Hydrogen Alpha (Ha for short). This is absolutely not true. (See the next figure.)

Ha emits light at 656nm (Ha is the primary emission type from Hydrogen, the most common element in the Universe, and a primary component in most nebulae) in the visible spectrum and a standard DSLR will capture it just as well as anything else in the visible spectrum. The problem comes because there are some very dim Ha nebulas which are just too dim to do easily with a standard DSLR. You can then "mod" your DSLR and have the UV/IR filter removed which has the benefit of increasing the red sensitivity substantially.

Keep in mind that when you do this, it massively increases the red in your images making everything very very red (including things that should not be red) and totally useless for normal photography without additional filters or heavily tweaking the custom white balance feature of your camera. Both Nikon and Canon cameras can be modded. I am currently still using an unmodded camera.

NGC2244 shot using a Hydrogen Alpha filter and an unmodded DSLR with the same exposure settings as would be used without the filter.

Regardless of the type of camera you choose to use you will need to connect it to your scope at some point.

For a DSLR you will need: a T-Ring which attaches to your camera and an adapter which connects your T-Ring to the scope. This adapter may be a prime focus adapter (if you can use one with your Dob) or a universal adapter. The difference is that a prime focus adapter allows a clear path from the telescope directly onto the sensor of your camera while the universal typically allows you to put an eyepiece inside the adapter to accomplish eyepiece projection.

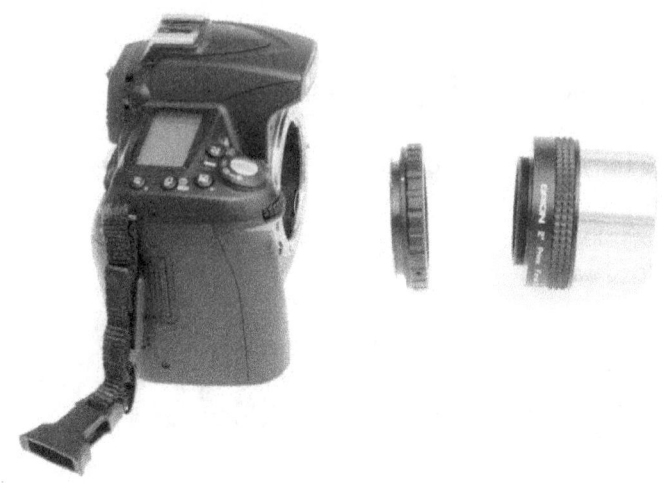

Camera, T-Ring and 2" prime focus adapter.

For DSLR work many people use the standard prime focus adapter. This adapter has a snout on one end that slips into the focuser of your scope while the other end threads onto the T-Ring. The T-Ring then attaches to your camera just like a lens would. Be careful, the T-Ring will often go on in several different positions, only one of which will lock in place. Make absolutely sure your T-Ring is fully locked on to the camera before you let go of it.

I have found only one reason for doing eyepiece projection. That would be if I were to use a telescope which was not designed for AP (such as a standard Newtonian, Dobsonians, or a Coronado PST). Such a telescope may not achieve focus with a prime focus adapter but might be able to using eyepiece projection.

The downside to eyepiece projection is that the resulting image is obviously inferior to what a prime focus system would produce. The eyepiece was never designed to project an image onto a camera sensor. It will however sometimes work, especially with something in the 20-26mm eyepiece range. Note that with these adapters you will be forced into using a physically very small eyepiece, usually a Plossl design. Use a very nice Tele-Vue Plossl ($120 for a 25mm) or at least an Orion HighLight Plossl ($60 for a 25mm) for best results.

Camera, 25mm Plossl eyepiece and eyepiece projection adapter.

Another little trick for scopes that will not come into focus is to use a Barlow. This will reduce the field of view quite a bit but it can indeed get images when nothing else will work. Again, when imaging optical quality is very important, a $125 Tele-Vue 2x barlow will far outperform a standard $45 2x barlow.

If none of this appeals to you and you cannot get your camera into focus you can try a low profile prime focus adapter, or replace your focuser with a low profile focuser.

The last thing I want to touch on, but by no means the least important, is how the camera's sensor relates to the field of view.

Larger sensors are like larger numbers on your eyepieces, they show a larger area of the sky. Smaller sensors are like smaller numbers on your eyepieces, showing a smaller area of the sky. For example, using a DSLR with a crop sensor (such as an APS-C size, most low to midrange DSLRs) you are "zoomed in" to your target more than if you use a full frame sensor (found in most high end DSLRs).

Smaller fields of view, more magnification, more zoomed in, whatever you want to call it can be useful for smaller targets but make larger targets much more difficult to image.

What I suggest is that you use a free program called Stellarium and the plugin that comes with it called Oculars. This will allow you to input your telescope focal length and sensor size to see what different targets would look like with that particular field of view. I have an introduction to Stellarium including the Oculars plugin in section 3.

Field rotation

One thing we have to work around is field rotation. The earth rotates around an axis, this axis is like a line going through the earth from the north to south poles and extending out into space. The point in space seems to be at different places depending on where you are on Earth.

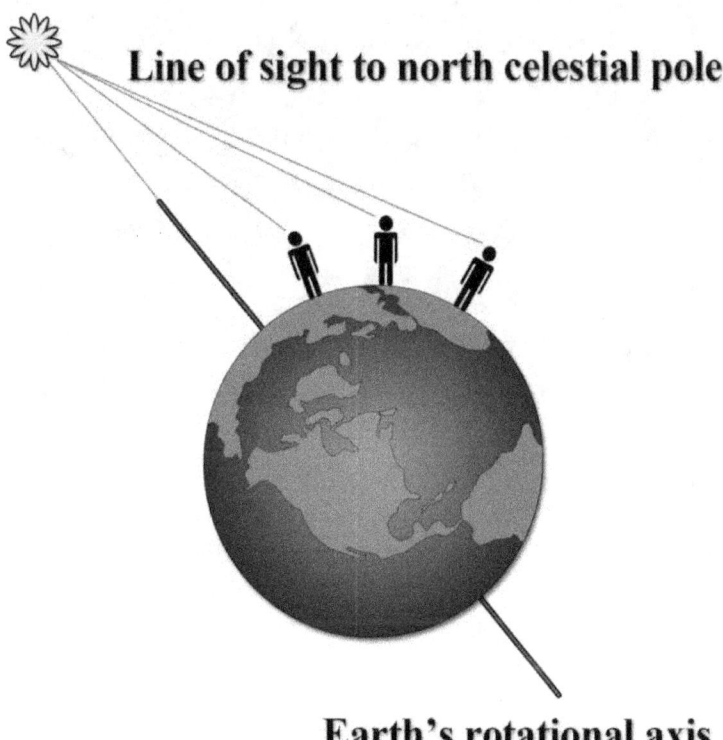

Line of sight to north celestial pole

Earth's rotational axis

Illustration of the Earth's rotational axis.

In the previous figure the star in the upper left corner represents the North Star, Polaris, the star that is almost exactly in line with the center of the Earth's axis of rotation. When looking at the night sky (or in the day for that matter) the sky seems to rotate around this point in space in the northern hemisphere. There is a similar point in the southern hemisphere.

129

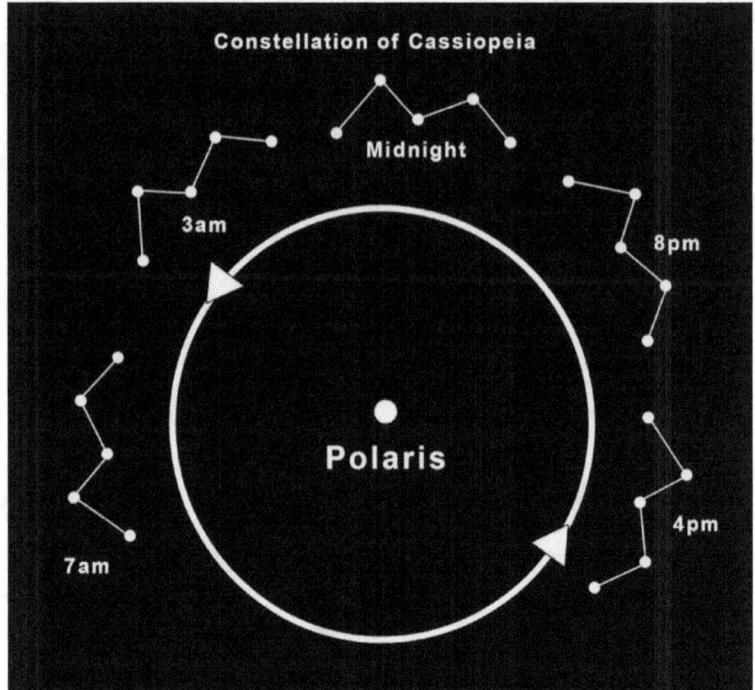

Illustration of how the stars circle the celestial pole.

In the previous figure note that the constellation Cassiopeia not only moves across the sky, but rotates as it does so. This motion is centered in the northern hemisphere on the North Star, Polaris. The center of the "W" always points towards Polaris.

What this means for you is that as an object you may want to photograph is moving across the sky, it is also rotating, so you need to account for both movements in order to take longer exposures or what you will wind up with will look like the following figure:

Illustration of unwanted field rotation.

The previous figure is representative of what would happen if you use what is called an Alt-Az (short for Altitude Azimuth, an inexpensive telescope mount that moves up, down, left and right) telescope mount, or a tripod, etc, to take images of a celestial object for a period of time greater than a few seconds.

In this case you are seeing the results on a galaxy, but the exact same problem occurs on any object when you use longer exposures, even the Sun.

What this means is that without an equatorial mount (which a Dob does not have), you are limited to a maximum of about 30 second exposures. In many cases, less. The greater the magnification, the shorter the exposure before the stars start to blur.

Fortunately, with digital cameras it is easy to test. Shoot one picture for 30 seconds and see if the stars are blurred. If so, reduce the exposure by 5 seconds and repeat. Keep going if necessary until you get acceptably round sharp stars.

Can you put the tube from a Dobsonian telescope on an equatorial mount? Technically you can, but there are problems.

The first way to accomplish this that everyone always thinks of is to simply strap their telescope tube to a big equatorial mount. The problem with this is that the tube is so large that the breeze from a passing fly at a hundred meters will cause the scope to move.

OK, maybe I was exaggerating a little bit on that last one. But any wind at all and it will move horribly.

In addition, the size of the tube makes it virtually impossible for all but the largest equatorial mounts (want to spend $10,000 on a mount, sure it will do it!) to track objects smoothly and reliably.

The correct way is to buy an equatorial platform to set your Dob on, base and all. You can buy these from such places as www.equatorialplatforms.com, or build your own with plans from places such as www.reinervogel.net/index_e.html.

While an equatorial platform will certainly extend your capabilities, they will never equal the astrophotography abilities of a good refractor on an equatorial mount. The other side of that is that the refractor on an EQ mount will never match the ease of use and simplicity of a good Dob.

Section 4: Intimate details

While it is true that the Dobsonian telescope is the simplest setup you can get, that is not to say there is nothing technical about them.

At some point you may want to know how to remove your focuser, or what exactly the primary mirror cell looks like, or even how twisting a screw moves where the secondary mirror points.

This is the section where we take things apart, measure them, and learn all the things we did not really want to know.

Complete disassembly

In this section I am going to completely disassemble the Orion Skyquest 8" Dob seen above. The purpose is to show you how one of these telescopes goes together.

While each manufacturer, and even each model from the same manufacturer, may differ substantially from each other, the basic premise behind the design is the same. This should allow you to take what you see here and apply it to any Dob made with minimal effort.

The nice thing here is that since you are watching me tear this telescope into tiny little pieces, there is absolutely no danger to your telescope. You get the benefit without the risk. How cool is that?

I have to mention that I do not recommend you do this. Getting your scope back to good working order after completely disassembling it can be difficult, particularly if you are not experienced with all aspects of collimation.

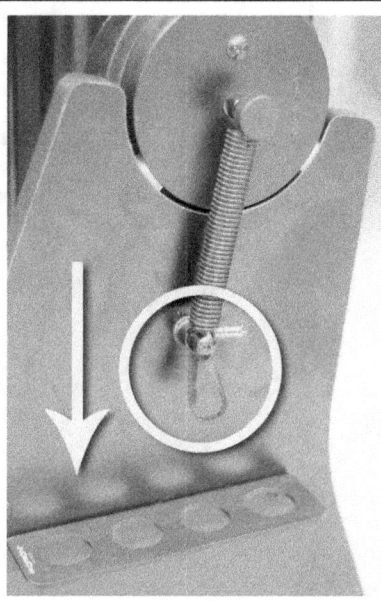

First, we need to take the telescope tube off the base. With this model, all we have to do is grab the cloth straps on the bottom of the spring on each side, and then we pull down lifting the hoop on the end of the spring over the bolt.

Once the springs are released, we simply lift the telescope tube off the base and set it on the floor.

When the telescope is sitting on the floor without a base it is extremely easy to knock over. Something like you bumping into it or a cat rubbing up against it is all that is needed to send it crashing to the floor. This can destroy the finder and even the secondary mirror assembly.

I never leave mine off the base for any length of time, and usually lay it down when I do so to make sure it does not get damaged. Tripping over the tube when it laying on the floor is a lot less likely to do serious damage than with it sitting upright balancing on just the tips of the adjustment screws on the base.

We remove the primary mirror cell now so that if we drop something inside the tube later, it does not fall and damage the mirror.

To remove the primary mirror cell you need to remove the six screws around the base of the telescope tube.

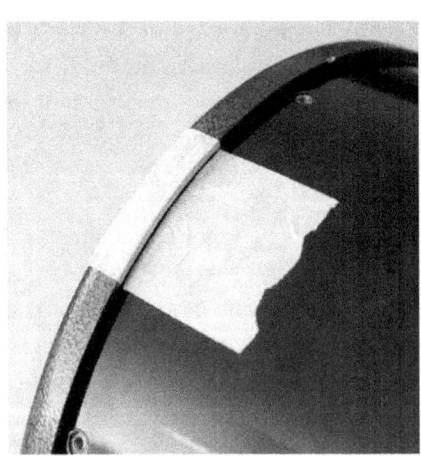

Now we mark the tube and mirror cell by using a piece of masking tape. This makes sure we get it back on in the correct position.

Once these screws are removed, you grab the ring at the bottom and pull it loose from the tube.

136

Be careful pulling the mirror cell as it is usually tight in the tube and is the single heaviest component in the telescope. This means if you are not very careful, when it lets go it can go flying across the room and destroy the mirror.

If possible, it is best to have one person holding the telescope tube while you use both hands to ease the mirror cell out of the tube by rocking it back and forth in the tube until one side comes loose.

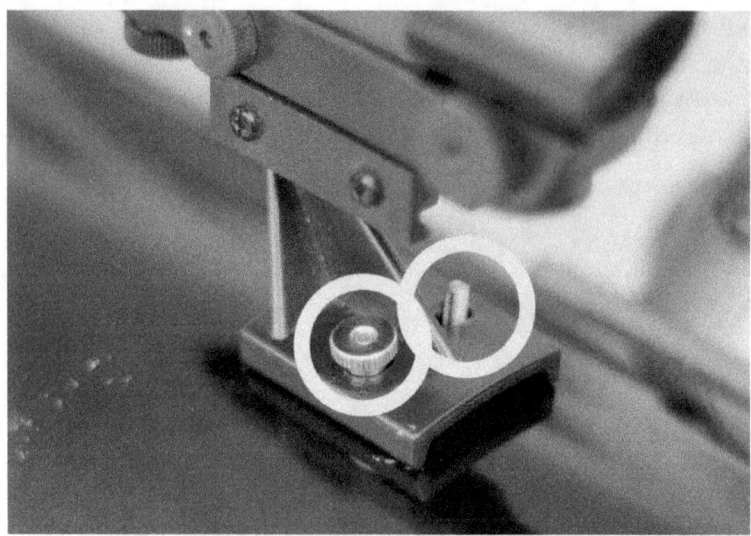

To remove the red dot finder you twist the two thumbscrews and remove them. Now lift the finder off the tube. The screws that the finder sits on have phillips heads inside the telescope tube and nuts on the outside of the tub that were revealed once you removed the finder.

Use a pair of pliers to hold the nut on the outside of the tube as you loosen the screws from the inside of the tube.

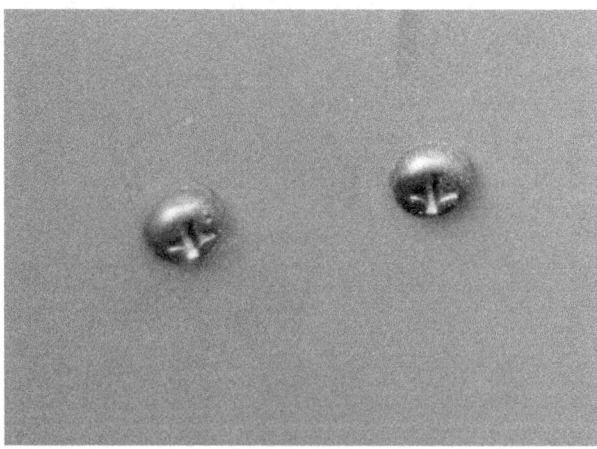

Working on the other end of the telescope we find the secondary mirror assembly being held in place by the three spider vanes. These vanes are metal bars being held on one end by screwing into the central mirror holding assembly, and on the other end by a screw and bolt setup like this:

The purpose of this setup is so you can adjust the length of the vane on any side and therefore move the central support to make sure it is centered where it needs to be.

In order to remove the end of the vanes you need to hold the nut on the inside of the tube and twist the thumbscrew (which is usually tighter than a thumb will loosen) on the outside of the tube.

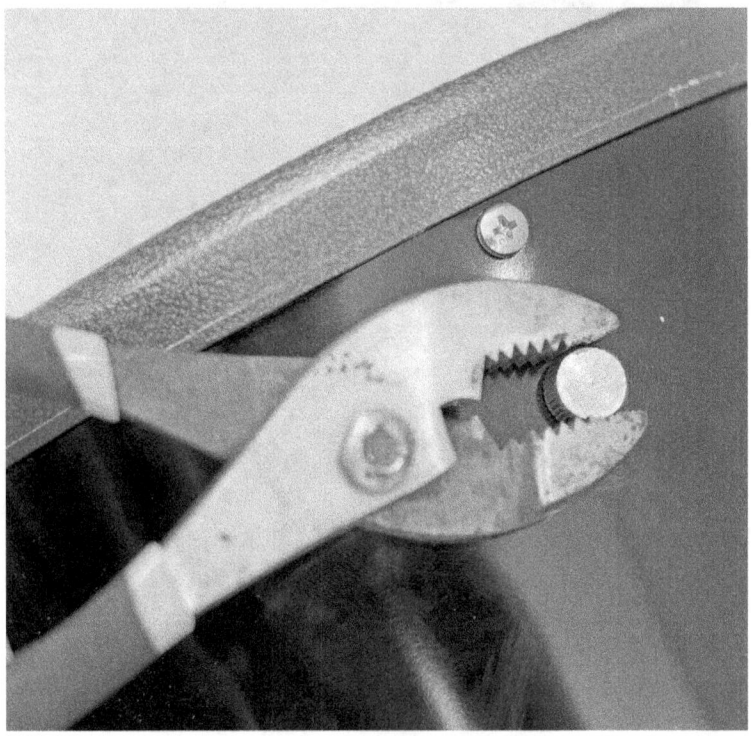

Once the thumbscrew portion comes loose, the nut on the inside will be free to move, and fall off, so be careful not to lose it.

When taking the last vane loose, you need to support the central section while removing the thumbscrew. Failure to do this may allow the mirror assembly to fall out and damage the mirror.

Also note the orientation of the mirror when removing it, it will need to be put back in facing the same direction.

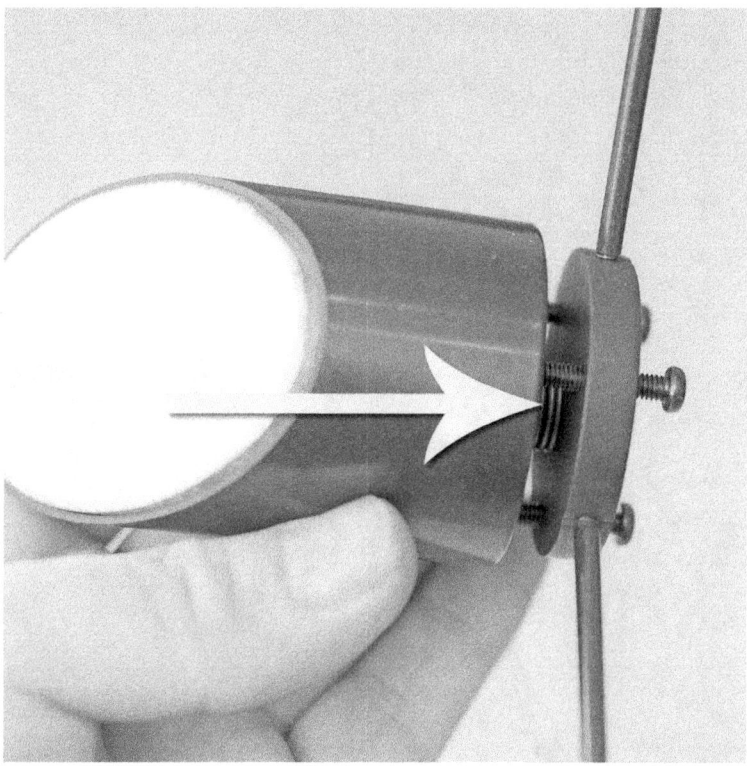

The secondary mirror assembly is held to the central support and vanes with a screw and spring. You can see the spring in the above image.

The three smaller screws are for adjusting the mirror.

In the image above, note the central screw which will detach the mirror assembly from the support section.

Also note the threads on the ends of the vanes as they thread into the central support.

In the above image the central support is detached from the secondary mirror assembly.

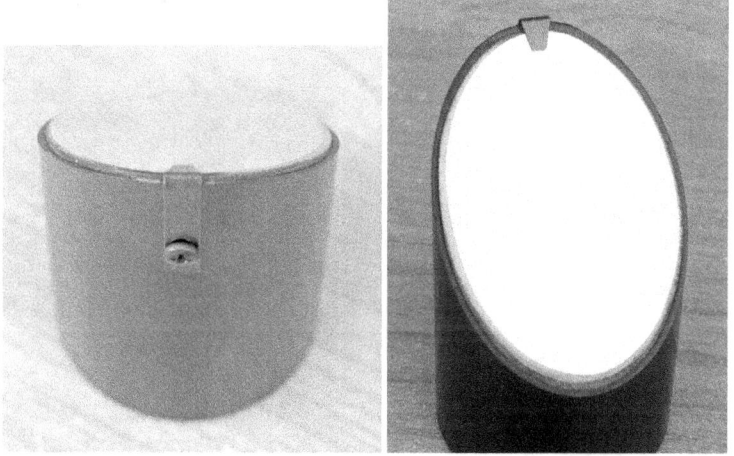

The secondary mirror is held into the mirror cell with one screw which holds a clip. This clip hooks the top of the mirror holding it down inside the cell.

Once the screw is removed the retaining clip falls off allowing the mirror to be removed from the mirror cell. Note the foam backing in the cell that the mirror sits on.

Removing the focuser is simply a matter of using a phillips screwdriver on the outside while holding the nuts on the inside. In my case these nuts did not require tools to hold.

Remember that the focuser has a little weight to it so once you take that last screw loose it will try to fall out.

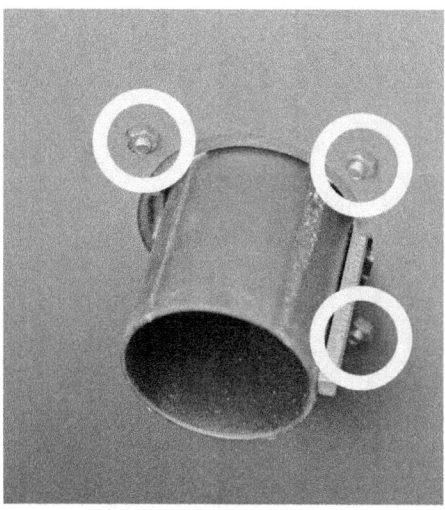

Once the focuser is out it becomes pretty easy to see this is a simple device that can readily be upgraded.

The hole where the focuser goes is shown below.

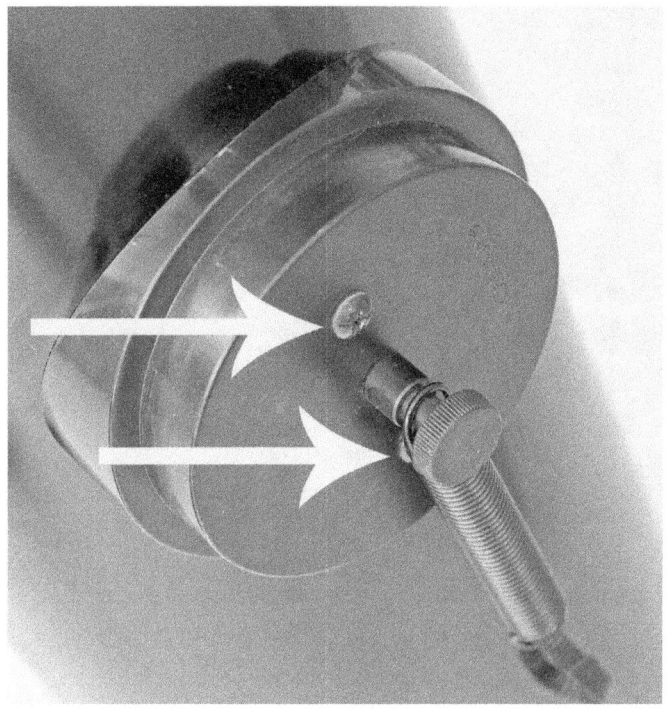

On each side of the telescope tube are the supports. Two screws run inside to nuts and washers on the inside. Holding these nuts by hand was easy however, yours may require holding them with pliers while loosening the screws on the outside.

Here is the back side of the supports with the washers and nuts.

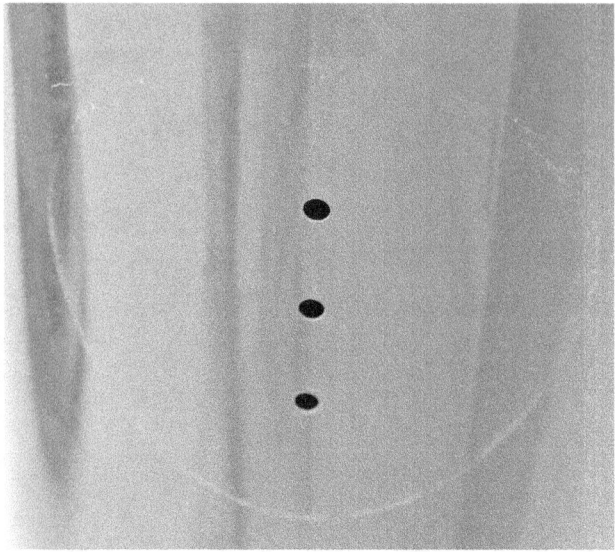

The image above shows the holes where the supports go.

The bolt that screws into the center of the support holds the tension spring that connects to the base. We released this at the beginning of this section.

On the top of the tube is the ring shown above which is used for decoration, to protect the end of the tube, and to attach the telescope tube cover to.

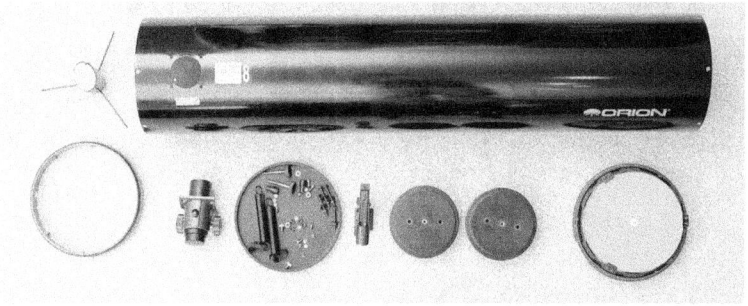

Here is the basic breakdown of the telescope into its major components.

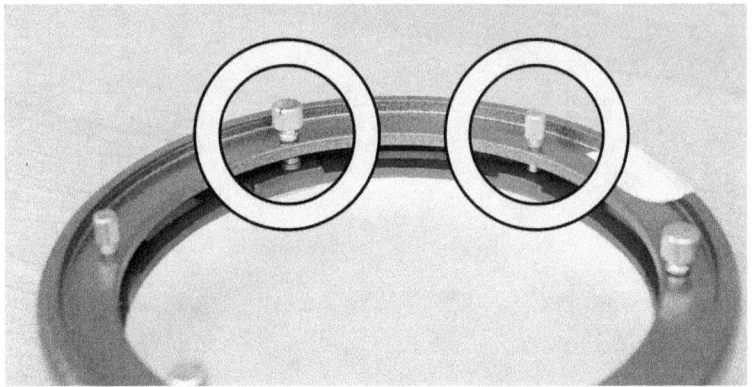

Disassembling the primary mirror cell is usually not advised. Unless you have a pressing need to do this, please do not. You will wind up causing yourself more headaches than you would think. You could even wind up causing severe damage.

There are two types of screws here, one set with springs under them which hold tension pushing out, and one set without springs that pull in. These two sets in opposition allow you to tilt the mirror into alignment.

Once we mark the assemblies so that we can put it back together just like it came apart, again using masking tape, we can remove all six screws which gives us the image above.

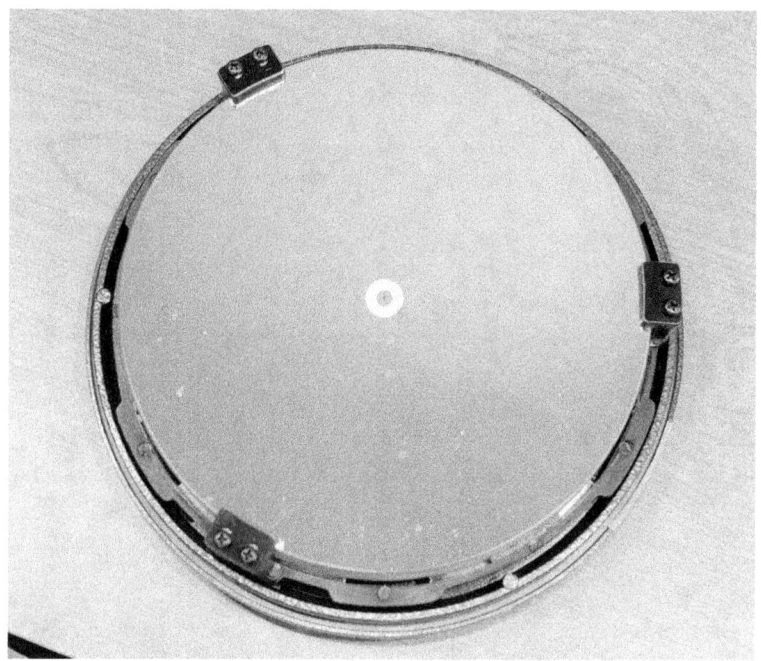

Look close at this mirror at how it looks so dirty. This is actually a pretty clean mirror and should not be cleaned. In almost all cases, people do more harm than good trying to clean a mirror. If you can read a printed page using the reflection, it is clean enough.

If you must clean your mirror, I suggest you remove it from the cell and put it in your sink on a folded towel for a cushion, then using just the water pressure from the faucet rinse the mirror off with cool water. Once done, rinse it again with pure isopropyl alcohol (I have had good luck with anything over 90% but use the highest you can get). Do not rub, wipe, or in any other way contact the mirror.

If this does not work and you still have dirt on your mirror, you can make sure the mirror is wet with alcohol, then use sterilized cotton balls to gently, and I mean gently, wipe across the surface. Do not press down or apply any force other than the weight of the cotton. Turn the cotton as you wipe and throw the cotton away as it becomes completely wet with alcohol.

Now let us continue with the disassembly.

There are three brackets with two screws each holding the mirror to the backing.

Note the cork pieces used to keep the glass off the backing acting as an insulator. Do not lose or damage these pieces!

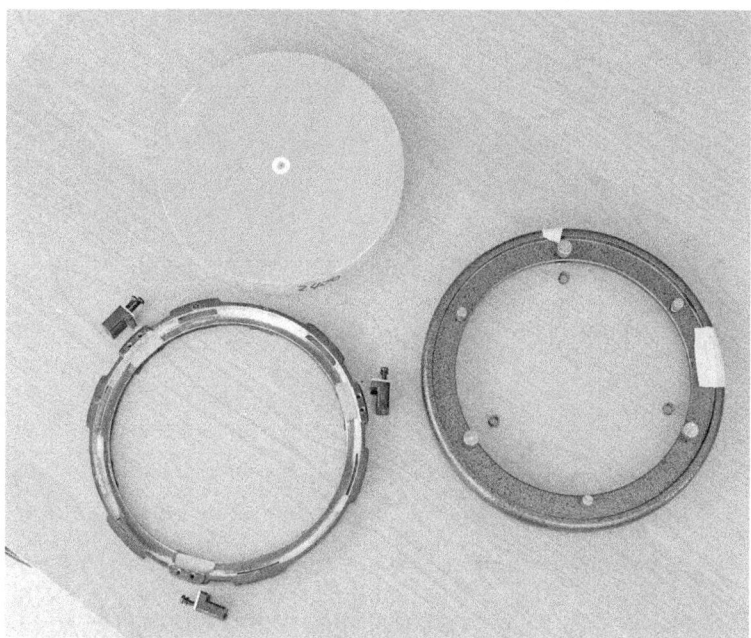

The completely disassembled primary mirror cell is shown above. Assuming you get everything back together correctly, you will still have to do a rather lengthy mirror alignment called collimation. Fortunately, that is covered in this book so you should be good.

The most difficult part of reassembly is the secondary mirror. Getting this spaced, pointed and angled correctly can be a real challenge.

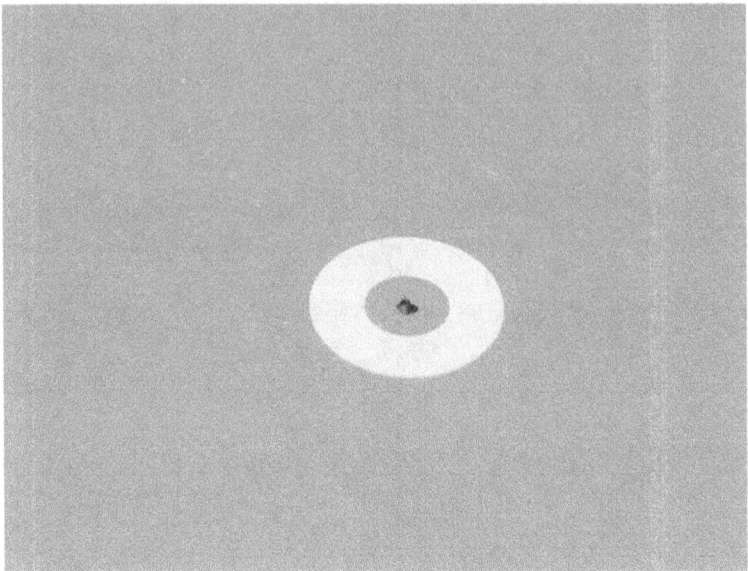

One last note is to look at the center of this mirror. The dot in the center is the factory center mark. While helpful, it is almost impossible to see when doing a collimation.

The white you see making a ring around the center dot is a hole protector which is available from any office supply store. They are used to repair a piece of loose leaf paper with holes in one side for putting in a three ring binder.

You will see these hole protectors used often on these mirrors as they are a perfect way to mark the center of the mirror. No, they are not visible when using the scope and cause no image degradation.

Average weights and sizes

The tables below are meant to help you get an idea of what will be practical for you to handle. Different manufacturers will have different weights depending on materials used and equipment installed (like focusers and finders).

Tube type:

Diameter	Weight lbs	Height inches
4.5"	18	39
6"	35	49
8"	50	50
10"	54	52
12"	80	62

Truss type:

Diameter	Weight lbs	Height inches
12"	84	62
14"	120	65
16"	195	76

Useful formulas

Figuring light gathering power:

Comparison ratio = larger aperture squared / smaller aperture squared

8" gathers 1.77 times more light than a 6"   1.77=64/36

12" gathers 4.5 times more light than a 6"   4.5=144/32

Magnification of a telescope and eyepiece combination:

Magnification = telescope focal length in mm / eyepiece in mm

A telescope with a 700mm focal length and a 10mm eyepiece gives us 70x:

70=700/10

Maximum magnification of a telescope:

Generally speaking, 2x the aperture. 100mm refractor = 200x. This varies wildly based on current seeing conditions and quality of the optics however this is a good maximum value for perfect conditions.

Section 5: Going further

The following sections are to help you get more out of this book, and to get more out of the hobby.

No book could ever contain everything you need for any hobby as complex astronomy, not even for a beginner. Hopefully the following sections will give you a good launch pad to extend your knowledge, and your enjoyment, of this awesome hobby.

Of course, the next logical step when you are ready is my book *Getting Started: Visual Astronomy*. I have some more detail on this book and a link to it in a later section titled "Other works by the author".

Where to get more information

Oh boy, are there a lot of places you can go to learn more, so here are some suggestions:

Astronomy equipment:
Orion Telescopes -www.telescope.com-1-800-447-1001
Agena Astro -www.agenaastro.com-1-562-215-4473
Oceanside Telescope -www.optcorp.com-1-800-483-6287
Shoestring Astronomy -www.shoestringastronomy.com
Astromart (used) -www.astromart.com
ScopeStuff -www.scopestuff.com

Online forums:
Astronomy Magazine -www.astronomy.com
Stargazers Lounge -www.stargazerslounge.com
Cloudy Nights -www.cloudynights.com
Ice In Space -www.iceinspace.com
CosmoQuest -www.cosmoquest.org/forum
Astromart -www.astromart.com/forums/

Specializations:
Spectroscopy -www.rspec-astro.com
Radio Astronomy -www.radio-astronomy.com
Photometry -www.citizensky.org
Astrophotography -www.allans-stuff.com/
-www.allans-stuff.com/youtube

Planetarium software:
TheSkyX -www.bisque.com
Starry Night -www.starrynight.com
Stellarium -www.stellarium.org
Cartes du Ciel -www.ap-i.net
C2A -www.astrosurf.com

Session planning software:
Astroplanner -www.astroplanner.net
Skytools -www.skyhound.com
Deep Sky Planner -www.knightware.biz

Index

Glossary

A/D converter (ADC) - Analog to digital converter. A camera sensor records light as an analog signal which the A/D converter then converts into digital information.

Achromat – A type of refractor typically with two lens elements to correct for chromatic aberrations. This type of scope is not well suited for astrophotography.

Afocal - A means of taking an image through an eyepiece of a telescope without removing the lens from the camera.

Alt/Az - Altitude Azimuth, a type of telescope mount that moves up and down, left and right as opposed to the smooth rolling motion of an EQ mount which accurately tracks the motion of the stars around the earth.

Amp glow – Amp glow is the glow that some cameras show on a long exposure image. This usually manifests itself in the corners of the image first and then can spread towards the center. A moderate amount of this can be removed using dark frames. Severe cases cannot be corrected.

Aperture - In telescopes, the diameter of the opening at the front of a telescope, usually measured in millimeters. Can also be measured in inches for larger scopes. In camera lenses there is a diaphragm inside the lens that controls the aperture which is sometimes referred to as an F-Stop.

Apochromatic (APO) – A type of refractor extremely well adjusted to remove most or all chromatic aberrations which makes it excellent for astrophotography uses. Can have two, three, or more lens elements. Higher end versions almost always have three or more elements.

Arc Minute – There are 360 degrees in the sky as it goes 360 degrees around us. One arc minute is $1/60^{th}$ of a degree.

Arc Second – Is equal to 1/60th of an arc minute.

Artifacts - Errors or unwanted signals in the image.

ASCOM - abbreviation for AStronomy Common Object Model and is a standard in the astronomy equipment industry for control interface design of astronomical equipment such as mounts, focusers, motorized domes, etc.

Astrograph - A type of Newtonian telescope that is designed specifically for astrophotography.

Astrometry – Extremely precise measuring of objects like comets and asteroids.

Astrophotography - Photography of objects in the sky.

Autoguider - A camera and associated equipment used to increase the accuracy of the mount in tracking the stars.

Audio Video Interleave (AVI) – A wrapper for computer video files, can contain a variety of different formats, typically video for Windows formats, and has a file extension of .AVI.

Back Focus – The necessary distance needed to be able to attach a camera onto a telescope focuser, and be able to bring the image projected onto that camera's sensor into focus.

Backlash – Unwanted spacing between gear assemblies usually resulting in some "play" or "slop" with the device. This is normally used to describe issues with a mount but can be applied to anything with gears.

Baffles – Ridges running around the inside of the light path in a telescope to prevent the scatter of light inside the telescope and provide an image with greater contrast.

Bahtinov mask - A mask or cover that goes in front of a telescope with a specific pattern of slits designed to provide easy focusing of point light sources such as stars.

Barlow - An optical device that increases the magnification or reduces the field of view, depending on how you look at it. This trades some image quality and light for more magnification. These plug into the optical train just before the eyepiece.

Bayer matrix - In color one shot cameras (any camera that produces a single color image in one exposure) the pixels are grouped in groups of four, one red, one blue and two green. These are combined to generate the color information for that area of the image. The Bayer matrix is the array of colored filters over the pixels that accomplishes this.

BFA – Bayer Filter Array, see Bayer matrix above.

Bias frame - An image taken with the highest shutter speed possible on a given camera at the same ISO and temperature of the light frames. This is used to subtract the camera's electrical signal present in every frame it takes from the final image.

Binning – A process of combining multiple pixels in order to boost sensor sensitivity at the expense of resolution. For example, 1x1 binning means each pixel counts as one pixel and is in effect not binned, 2x2 binning would take a square of 4 pixels and combine them into one "super pixel".

Binos - Short for binoculars.

Bino-Viewer - A device that allows attaching two eyepieces to a standard telescope so you may view objects in stereo.

Bit - A single bit can be either on or off, representing either 0 or 1. Computers use this as the basic language of everything they do.

Bit depth - This describes a measurement of something like the number of colors an image can contain and is base two mathematics. An example is a 1 bit scale will contain two possible combinations, a 2 bit scale will contain 4, a 4 bit scale will contain 16 and an 8 bit scale will contain 256 bits.

Black point - An area of an image that represents absolute black.

Blooming - In a camera, once a pixel has received as much light as it can handle, the voltage can spill over into adjacent pixels causing them to be brighter than they should.

Bortle scale – Astronomer John Bortle developed a scale of nine levels which represents the "true darkness" of a site, or the amount of light pollution present.

Bulb exposure – A bulb exposure is an exposure where as long as the shutter button is held the camera continues the exposure. DSLRs and other cameras can be used in this mode.

CCD - Short for Charged-Coupled Device, a type of sensor used in digital cameras. In astrophotography it is usually used as a reference to a camera designed and used specifically for astrophotography as opposed to a digital SLR or other multi use digital camera.

Celestial equator - An imaginary line which is basically the equator of the earth projected up into the sky.

Center mark – A dot placed exactly in the center of the primary mirror of a Newtonian to aid in collimation.

Chromatic aberration – Chromatic aberration is the "glowing" or "fringing" of light around bright objects in a telescope. This is caused when light passes through the optical path it is split into its component colors and then rejoined imperfectly at the focal point.

Clip - Clipping an image means you have cut off one end or the other of the image's ability to record data (as can be shown in a histogram). Clipping the highlights for example means that area of the image is pure white and cannot contain any detail. Clipping the darks means that part of the image is pure black and contains no detail.

CMOS - Complimentary Metal Oxide Semiconductor. In astrophotography, a type of sensor in a camera.

Collimation - The act of aligning the optical components of a telescope to make sure all parts of an image combine correctly into one sharp image.

Coma - An optical defect normally present in reflector telescopes that can cause point light sources such as stars to appear to be out of round, presenting like they have the tail of a comet.

Coma corrector - An optical device for reflector telescopes to correct for coma aberrations.

Convolution – A mathematical method of multiplying arrays of numbers to get a third array of numbers. Used in image processing to stretch or resize images.

Corrector plate – The lens on the front of an SCT type telescope that corrects for the spherical aberration created by the spherical mirrors used in that design.

Counterweight – A weight, usually on an equatorial mount, used to balance the weight of the telescope and associated hardware.

Crayford focuser – A telescope focuser that uses smooth bearings and rollers as opposed to gears used in rack and pinion style. They usually come in dual speed (coarse and fine adjustments) and can have adjustable tension.

CRW/CR2 - Canon's RAW image format.

Dark frame - An image taken at the same ISO, shutter speed and temperature as the light frames but with the lens cap/scope cap on, or the shutter closed. This is used to detect the thermal signature of the camera's sensor at these setting so they can be subtracted from your final image.

Dead pixel - Opposite of a hot pixel, a pixel that is stuck in the off position and registers as black regardless of the amount of light applied.

Declination (DEC) – Celestial coordinate measured from the celestial equator north and south of that line, from +90 degrees to the north to -90 degrees to the south, zero being the celestial equator.

Deconvolution - A method of image enhancement that corrects for the bad effects of convolution. This can substantially increase fine details in an image.

Dew heater - Usually a strip that heats up and is wrapped around a telescope near the optics. This warms the optics and prevents dew from forming.

Dew shield - A device attached to the end of a telescope and is like a hollow extension of the telescope tube. This delays the objective from collecting dew, and reduces the intake of extraneous light sources.

Diagonal - A device that has a mirror inside and reflects the image at a 45 degree or 90 degree angle for easier viewing. One side goes into the focuser, the other end holds an eyepiece.

Diffraction - As light passes through a telescope it passes through openings. As light gets near the edges of these openings it is diffracted. This causes stars to appear larger than they actually should.

Diffraction limited – Term used primarily by telescope manufacturers that says that the telescope should perform so that any defect seen will be with the physical characteristics of light and not optical problems with the telescope.

Dispersion – Cause of chromatic aberrations. Prism effect, when light is spread out into its spectrum from white light.

Dobsonian - a type of telescope mount, but usually used as a reference to the entire telescope assembly. These are usually larger Newtonians mounted onto a base that sits on the ground and moves as an alt/az. Like regular Newtonians these are not well suited to astrophotography due to not having enough backfocus.

Doublet – A refractor telescope with two objective lenses.

Dovetail - A metal rail that attaches to the bottom of the telescope, usually by rings that clamp into the telescope tubes or bolts into the bottom of the telescope, which can then be quickly and easily attached to the mount's clamp. Popular dovetail types include Vixen and Losmandy.

DSLR - Digital Single Lens Reflex camera. A type of camera where the user actually looks at the same image that will be recorded on the sensor by means of a mirror and prism that reflects the light from the lens through an eyepiece. When the shutter is opened to take the picture the mirror swings out of the way, the eyepiece goes black as it is no longer receiving the reflected image, and the sensor is exposed.

DSS – Short for Deep Sky Stacker, very popular free program generally used by beginning astrophotographers for stacking images.

Dynamic range - The range from brightest to darkest that a camera can record.

ED – Extra low Dispersion, optical glass corrected for chromatic aberration.

EQ/Equatorial Mount - A type of mount specifically designed to track the stars as they travel around the earth compensating perfectly for their arc in the sky.

Ephemeris – Detailed positional information about planets, their moons, comets and asteroids.

Eyepiece - An optical device that focuses the light exiting a telescope tube in such a way that you can view it with your eye. These typically contain many lens elements in a round cylinder that is inserted into the focuser. The eyepiece can be made to magnify or reduce the image size.

Eyepiece projection - A method of taking a photograph through the eyepiece of a telescope without a lens on your camera. This uses a specific adapter. This can come in handy on telescopes that cannot reach focus using a prime focus adapter.

F-Stop - When using a camera with its lens installed, the aperture is adjustable and is commonly referred to as the F-Stop.

Field flattener - An optical device used primarily on refractors to make sure that the image arrives at the camera sensor perfectly flat. This prevents elliptical images of stars in the corners of the images while the stars in the center may be perfectly round.

Field of view - Commonly represented as FOV. The area of the sky that you can see at one time. Longer focal lengths (more magnification) generally show smaller areas of the sky and hence a smaller field of view. Eyepieces with smaller numbers cause the same effect.

Field rotation – The effect of the image being blurred from the rotation of the sky. This can happen when you use an Alt/Az mount to take long exposures since the Alt/Az mount does not rotate the camera like an EQ mount does.

Filter – A filter is a piece of glass (or Mylar in some solar filters) that alters the light coming through the telescope before the eyepiece or camera. A filter is used for removing light pollution, enhancing certain colors, shooting color images with a monochrome camera and many other tasks.

Finder - A small telescope or other pointing device that helps you quickly orient your telescope towards a particular target. Similar to a gun sight.

Firmware - The software a device uses to tell it what to do. For example, your GoTo telescope software in the hand controller is called its firmware and can be updated on many devices.

FITS format - A file format designated by .FIT (such as .TIF, .GIF or .JPG) specifically designed for scientific purposes. Like RAW or TIF files this stores raw data that does not degrade from repeated editing as do formats such as .GIF or .JPG.

Flats/Flat frame - An image taken with even illumination over the front of the telescope and exposed to present a neutral gray image. This must be taken with the exact same setup as your light frames (same focus setting, same filters, etc) and is used to remove vignetting.

Focal length - The length of a line following where the light travels through a telescope, this is important for calculating parameters such as the FOV and magnification.

Focal plane – An inferred plane at the point where the image from the telescope comes to focus. A camera's sensor is mounted so that it is at the focal plane.

Focal ratio (FR) – The focal length divided by the aperture of the primary objective of the telescope.

Focal reducer - An optical device which reduces the effective focal length and increases the field of view of a telescope, seemingly reducing the magnification. This is usually mounted into the focuser before any eyepieces or cameras.

Focuser - A piece of equipment mounted on the telescope where the light exits. Eyepieces, diagonals, barlows and cameras are mounted into the focuser. Its job is to move the eyepiece/camera/etc back and forth until the light comes into focus at a specific point (your eye or the camera sensor).

FOV – See field of view.

Frames per second (FPS) – The number of image frames captured per second by the device, used in video capture devices.

Full well capacity - A measurement of the total amount of light a photosite can store before saturation occurs.

FWHM – Full Width Half Maximum. The measurement of the angular apparent size of a star, usually used to get the size as small as possible in an image which represents the best possible focus.

Gain - This is a multiplication of the incoming signal. For example, if one photon enters a camera and hits the sensor, setting the gain to 2x will cause the digital signal sent from the camera sensor to say that two photons hit the sensor. Increasing the ISO of a digital camera is increasing the gain.

German equatorial mount (GEM) – Another name for the equatorial mount.

GoTo – A telescope that when properly aligned can point to a celestial object automatically when selected from a catalog or menu.

GPS – Global positioning system, a device or feature used to determine your exact location on the planet.

Grayscale - An image recorded in black, white, and variations of gray with no color information.

Guiding - The act of following a star or other object using either manual corrections (as was the case back before GoTo and tracking mounts) or automatically using guiding equipment such as an autoguider.

Hand controller (HC) – The handheld device used to control your telescope's mount.

HDR - High Dynamic Range. You can use different exposures on different images and sandwich them together to show an image that has too much dynamic range to be captured in one single exposure. M42 is a prime example of a target that needs HDR processing: if you expose correctly for the faint dust lanes on the outer areas, the central core is blown out or clipped; if you expose for the central core, the outer dust lanes are clipped into blackness and can not be seen.

Highlights - Areas of maximum brightness in an image.

Histogram - A graph that shows how an image is exposed. In a normal grayscale histogram the left side is absolute black, the right side is absolute white and there is usually a hump in the graph display somewhere near the center showing the exposure of that image. Color works the same way but shows the intensity of the red, blue and green color channels.

Hot pixel - Opposite of a dead pixel. A pixel that shows exposure information even when shot in complete darkness.

Illuminated reticle eyepiece – An eyepiece with an illuminated crosshair or other centering marker used for precise centering of targets in the field of view.

ISO - International Standards Organization, used to measure the "speed" of film, or the sensitivity of a sensor in a digital camera. As ISO increases, less light is required to "expose" for a given image. This also reduces the signal to noise ratio, increases noise, and reduces the bit depth possible in the image.

JPG/JPEG - Joint Photographic Experts Group. A file format denoted by .JPG (such as .TIF or .GIF) that is very common in digital cameras. Using this format should be avoided because it uses a lossy compression format to reduce file size. This results in huge losses of information and makes it virtually impossible to process well for astronomical uses.

Light frame - A standard picture. Every regular picture you have taken with a regular camera of birthdays, friends and family are all what we call light frames. These are the frames you work with that contain your image data.

Light pollution – Stray light from street lights, signs, windows etc that shine or are reflected up into the air. This is scattered by contaminates and humidity in the air and create a glow effect around cities making it difficult to see outside the atmosphere.

Light year – The distance light travels in a year through a vacuum, approximately 5.87 trillion miles.

Limiting magnitude – The measurement of the dimmest star you can see at zenith which takes into consideration all parameters such as light pollution, weather conditions and optical devices used (if any).

Lossless compression - Certain file formats such as PSD and TIF employ compression methods that preserve 100% of the data while decreasing the file size.

Lossy compression - Formats such as .GIF and .JPG use lossy compression which throws away data that it does not think is needed to display the image.

LRGB - When shooting a monochrome camera and creating a color image you need to shoot at least one image with a red filter, one image with a green filter and one image with a blue filter. These are combined together into one color image. The L in LRGB stands for luminance and is used to increase detail in an image. The Luminance frame is the detail frame and can be shot in very high resolution. The color can be shot at lower resolutions and combined with the luminance to create a high resolution color image. You can use this idea to increase your ability to stretch images as well.

Luminance – The recording of brightness or intensity of light. Typically this is the high resolution/detailed portion of an image.

Magnitude - A measurement of the brightness of an object. An increase in one magnitude is approximately 2.5 times as bright. The lower the number on the scale, the higher the magnitude.

Maksutov Cassegrain telescope – See MCT below.

Maksutov Newtonian – Similar to a Maksutov Cassegrain except they are designed as a Newtonian configuration with the focuser near the front of the scope.

MCT - Maksutov Cassegrain Telescope, a type of telescope that has a sealed front end which is actually a corrector lens called a meniscus, two mirrors and has its eyepiece in the rear.

Megapixel - Roughly one million pixels.

Meridian - An imaginary line dividing the west and east halves of the sky running from the north celestial pole directly overhead to the south celestial pole.

Meridian flip - Meridian Flip is the act of re-orienting the scope on an EQ mount so it can continue to track past the meridian. This "flips" the scope around to pointing the other direction at roughly the same spot on the meridian. Going past the meridian without flipping can cause the scope to run into the mount, cables to come loose, and many other really bad things.

Micron – One millionth of a meter or 0.001mm.

Mirror cell – The frame that holds the primary mirror assembly.

Mirror lock(DSLR) – Some cameras have the ability to lock the mirror in the up position to minimize camera vibration when the shutter is tripped. This can be very useful shooting brighter objects like the moon but is ignored in long exposure work as the amount of time the camera is vibrating due to the mirror slamming open is miniscule compared to the overall exposure time.

Mirror lock(SCT) – Some SCT type telescopes have the ability to lock the mirror once the image is in focus to prevent the mirror from "flopping" or moving as the orientation of the telescope changes.

Monochrome – Technically means one color, meaning either black or white. "Monochrome" cameras are actually grayscale in that they produce black, white and many different shades of gray.

Mosaic - The act of shooting multiple images in a grid pattern and stitching them together to allow you to shoot a larger field of view than you could normally.

Mount - The mount is the geared (and sometimes motorized) device that is typically attached to the top of a tripod and then has the telescope attached to it. It is the mount that allows you to point the telescope at different objects without moving the tripod, and (when motorized) tracks objects across the sky.

Narrowband - Using special filters you can capture the emissions from certain gasses such as hydrogen alpha, sulpher and oxygen. These can be used much like LRGB imaging to create faux color images of high resolution. This method can also overcome all but the worst light pollution situations and can even allow you to shoot on nights with a full moon to some degree.

Near Earth Object (NEO) – An object such as a comet or asteroid which will pass in close proximity to earth.

Newtonian - A type of reflector telescope that has two mirrors in a hollow tube. The front of the telescope is open to the elements and the back is sealed. The eyepiece is near the front of the scope. These are usually not suitable for astrophotography unless they are designed as an "astrograph" as they will not bring a camera to focus without modifications or the use of a Barlow.

North celestial pole (NCP) – The point in space very close to Polaris where a line drawn from the exact southern to northern poles would extend into space with the earth revolving around that line.

Nyquist theory - States that when converting frequencies, the sampling rate should be 2x the highest frequency to get an accurate conversion and preserve all the data.

Objective lens – Also called the primary objective, the large front lens of a refractor telescope.

Off axis guider (OAG) – A method of mounting a guide camera so that it shares most of the same optical path as the imager, picking off a small amount of light usually from a mirror mounted in the light path.

One shot color (OSC) - Any camera that creates a color image from a single exposure.

Opposition – Opposition is when a planet is closest to the earth and is directly on the other side of earth from the sun.

Optical train - Anything that is directly in the path of light from the stars to your eye or camera sensor is considered "in the optical train". Could be called the optical path as well.

Optical tube assembly (OTA) – Also referred to as the OTA, this is the main tube of the telescope not including any mount, pedestal, pier or tripod.

Parfocal – Applies to both eyepieces and filters and means that if you exchange one filter (or eyepiece) for another, you will remain in nearly perfect focus. Not all filter sets or eyepiece sets are parfocal.

Periodic error (PE) - Errors in the manufacturing process of the gears and drive assembly in an EQ telescope mount results in repeating errors in the tracking of the mount. These can be removed with software that contains PEC code.

PEC - Periodic Error Correction. Software that corrects for periodic error.

Photometry – The measurement of apparent magnitude of objects such as comets, asteroids and stars.

Photon - For the purposes of discussion in this book, a photon is a single particle of light.

Photosite - The technical name for the tiny part of the sensor in a digital camera sensor that when exposed to light records a signal. Typically called a pixel.

Piggyback - Mounting a camera with a lens on a telescope in such a way as it is not shooting through the telescope but is instead just using it as a tracking mount.

Pixel - A single dot in an image.

Pixel size - The physical size of a photosite on the sensor of a camera, measured in microns.

Plate solve – Refers to Plate Solution, or finding the absolute position and motion of an object. Some applications such as TheSkyX Professional offer a plate solve feature where it can look at your image and tell you exactly what is in the frame.

Point light source - Stars are considered point light sources because regardless of their magnification they are so far away they will always appear as a single point of light.

Polar alignment – Aligning the "polar axis" of an equatorial mount to either the northern or southern celestial pole so that the mount can track celestial objects precisely.

Polar scope - A small telescope usually built into the mount which allows for precise pointing of the mount's right ascension axis to the north or south celestial pole.

Prime focus - Attaching a camera without a lens in such a way that the image from the telescope is directly projected onto the sensor of the camera.

Quantum efficiency (QE) - A measurement of the percentage of photons which hit a photosite versus how many are detected.

Rack and pinion focuser – A less expensive and typically less accurate style of focuser.

RAW - A RAW file is a file that contains the relatively unaltered, unmodified data directly from the camera's sensor.

Rayleigh scattering – The scattering of different wavelengths of light by the molecules in the atmosphere. This scattering is the reason the sky appears blue.

Resolving power – 4.56/(inches of aperture of the telescope)=resolving power of the telescope in arc-sec. Note that this does not take into consideration obstructions such as secondary mirrors.

Reticle - Crosshairs or other markings that allow you to precisely center a target in your field of view. Sometimes included inside eyepieces and finder scopes.

Red dot finder - A type of finder that uses an illuminated red dot as a reticle.

Refractor - A type of telescope that has an objective lens on the front end and an eyepiece or camera at the other. Light passes straight through without being reflected unless a diagonal is used.

RGB - Red, Green, Blue. One shot color cameras shoot everything as a combination of these three primary colors. When shooting monochrome images and wanting to end up with a color image, you shoot at least one frame with a red filter, one with a green, and one with a blue and then combine them to create a full color image.

Right ascension (RA) – Celestial coordinate measured from west to east in hours, minutes and seconds. As the earth turns each hour, 15 degrees of arc pass.

Saturation – The point at which you cannot record any more data. This may refer to the full well capacity of a CCD camera or the maximum value a pixel can store.

Schmidt Cassegrain Telescope (SCT) - a type of reflector that has a sealed front, two mirrors and has its eyepiece in the rear of the scope.

Seeing - A measurement of the conditions of the atmosphere as it relates to being able to view or image an astronomical object. An easy method to determine the seeing conditions is to look for stars twinkling; the more they twinkle, the worse the seeing.

Sidereal rate – 23 hours, 56 minutes and 4 seconds is one sidereal day which is why the stars are never at the exact same place at the exact same time every night and seem to "advance" across the night sky every night all year long. This is the rate at which your telescope must track to remain aligned with your target.

Signal to noise ratio (SNR) - The ratio of signal (what you are trying to capture in the image) to noise (electrical signals inherent to the camera generating the image). The higher the SNR, the easier it is to stretch an image and bring out the detail of your target.

Slew – The process of your telescope moving to and from targets.

South celestial pole (SCP) – The point in space very close to Sigma Octantis where a line drawn from the exact northern to southern poles would extend into space with the earth revolving around that line.

Spider vanes - Small strips of metal or plastic in the front of a Newtonian telescope which supports the secondary mirror in the optical path.

Stacking - Taking several images and combining them in such a way as to increase the signal that you want to keep while reducing the noise levels that you do not.

Strehl ratio – Gives a ratio as compared to a theoretically perfect optical system. For example, a Strehl ratio of .90 is 90% as good as a theoretically perfect optical system.

Stretching - Taking an image and manipulating the data so that details that were too dark to see are now light enough to be visible through compression of the grayscale or color scale.

T-Ring – An adapter that mates with a removable lens camera on one side and has threads on the other side to attach to the telescope or other device.

Thermo Electric Cooler (TEC) – Electric cooling device used with some CCD and DSLR cameras.

TIF - A file type (like .GIF and .JPG) to store image files. TIFs are excellent because they are lossless formats. They are however far larger than JPG or GIFs.

Tracking - The ability to follow an object as it appears to travel across the sky.

TSX - Abbreviation for TheSkyX, a planetarium, telescope control and planning application for amateur and professional use from Software Bisque Inc.

United States Naval Observatory (USNO) – The standard for timekeeping in the United States.

Vignetting - The effect of the edges of an image being darker than the center due to obstructions or optical imperfections.

Well depth - A measurement of the total amount of light a photosite can store before saturation occurs.

White point - A part of an image that represents pure white.

Zenith – The point directly overhead.

Other books by the author:

Want to take a few snapshots of the beautiful objects you are viewing without spending a small fortune? Already have a camera but you can't seem to get a good image and want to know why?

This book will answer those and many other questions while giving you a quick and reasonably easy introduction to budget astrophotography. In addition, save more money by seeing how to make a lot of items you may find useful.

http://www.allans-stuff.com/bap/

If you decide that you want more than quick snapshots, you want big beautiful prints to hang on your wall, this is the book for you.

From required and optional equipment, through the capture process and into the software processing needed to create outstanding images, this book will walk you through it all.

http://www.allans-stuff.com/leap/

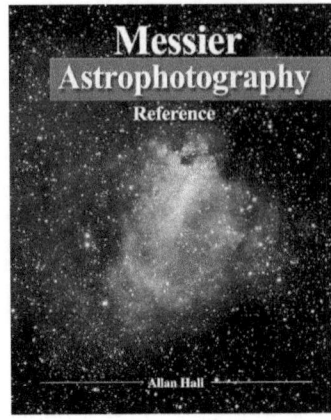

You decide that you want to take images of celestial targets, but need a little help with the targets? This book discusses all 110 Messier targets and includes descriptions, realistic images of each target, star charts and shoot notes to help you image all 110 of the objects yourself.

http://www.allans-stuff.com/mar/

If you have ever wanted to view the wondrous objects of our solar system and beyond, here is the how-to manual to get you well on your way. From purchasing your first telescope, through setting it up and finding objects, to viewing your first galaxy, this book contains everything you need. Learn how to read star maps and navigate the celestial sphere

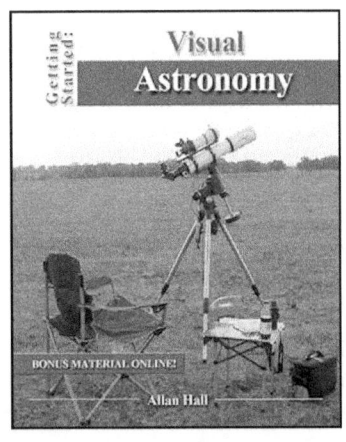

and much more with plenty of pictures, diagrams and charts to make it easy. Written specifically for the novice and assuming the reader has no knowledge of astronomy makes sure that all topics are explained thoroughly from the ground up. Use this book to embark on a fantastic new hobby and learn about the universe at the same time!

http://www.allans-stuff.com/va/

Many midrange and high-end telescopes come on equatorial mounts. These mounts are fantastic for tracking celestial objects. Someone who wanted to take pictures of objects in the night sky might even say they are required for all but the most basic astrophotography. The problem is that they can also be unintuitive and require some knowledge to use.

If you have ever struggled to figure out how to use an equatorial telescope mount, this is the book you always wished you had.

http://www.allans-stuff.com/eq/

Do you want to learn how to take photographs of an exciting Solar or Lunar Eclipse? Do you have the right equipment for the job? Do you want to know ALL the tips and techniques needed to make this a success?

Capturing these amazing, possibly once-in-a-generation events is something that you won't want to miss out on and capturing the best shots of it is crucial when it comes to the bragging rights. Get your copy of How to Take Pictures of an Eclipse now and make sure that you are ready to get the photographs that will amaze your friends and family and be the envy of all for years to come.

http://www.allans-stuff.com/eclipse/

So you've decided to write a book and get into non-fiction publishing. Now you find yourself faced with the seemingly infinitely harder second step – actually bringing the idea to market. In today's brave new world of self-publishing and open creative markets, it is both an inviting and potentially intimidating arena for authors hoping to turn their non-fiction books into a meaningful source of income. This is a daunting task because it involves a blend of several disciplines that aren't necessarily part of an author's quiver of arrows. Most crucial among these are marketing and digital publishing, each of which requires fluency in fields that authors may or may not have experience in.

http://www.allans-stuff.com/ck/

WordPress is the perfect tool to help you build the website you've always wanted. But the 'help' aspect which is built into it isn't always the right thing for someone who just getting started.

What you need, and what this book will provide, is a book that shows you how to get off the ground and then build on that knowledge to give you a secure and usable website.

http://www.allans-stuff.com/wp/

Information Technology is an area which is constantly on the move, sometimes at a speed which is dizzying and difficult to keep pace with. In particular **data recovery** can be one of the more complex problems you might encounter. The sheer amount of information is often overwhelming and confusing.

Data Recovery for Normal People is a new book which aims to make this process a lot simpler. Designed for both beginners who have little knowledge of technical issues and for those who may own their own computing business and want to learn more.

http://www.allans-stuff.com/dr/

Choosing and Using a Dobsonian Telescope

NOTES:

NOTES:

www.ingramcontent.com/pod-product-compliance
Lightning Source LLC
Chambersburg PA
CBHW070027210526
45170CB00012B/219